Advances on Models, Characterizations and Applications

STATISTICS: Textbooks and Monographs

Recent Titles

Handbook of Applied Economic Statistics, *Aman Ullah and David E. A. Giles*

Improving Efficiency by Shrinkage: The James-Stein and Ridge Regression Estimators, *Marvin H. J. Gruber*

Nonparametric Regression and Spline Smoothing: Second Edition, *Randall L. Eubank*

Asymptotics, Nonparametrics, and Time Series, *edited by Subir Ghosh*

Multivariate Analysis, Design of Experiments, and Survey Sampling, edited by Subir Ghosh

Statistical Process Monitoring and Control, *edited by Sung H. Park and G. Geoffrey Vining*

Statistics for the 21st Century: Methodologies for Applications of the Future, *edited by C. R. Rao and Gábor J. Székely*

Probability and Statistical Inference, *Nitis Mukhopadhyay*

Handbook of Stochastic Analysis and Applications, *edited by D. Kannan and V. Lakshmikantham*

Testing for Normality, *Henry C. Thode, Jr.*

Handbook of Applied Econometrics and Statistical Inference, *edited by Aman Ullah, Alan T. K. Wan, and Anoop Chaturvedi*

Visualizing Statistical Models and Concepts, *R. W. Farebrother and Michael Schyns*

Financial and Actuarial Statistics, *Dale Borowiak*

Nonparametric Statistical Inference, Fourth Edition, Revised and Expanded, *edited by Jean Dickinson Gibbons and Subhabrata Chakraborti*

Computer-Aided Econometrics, *edited by David EA. Giles*

The EM Algorithm and Related Statistical Models, *edited by Michiko Watanabe and Kazunori Yamaguchi*

Multivariate Statistical Analysis, Second Edition, Revised and Expanded, *Narayan C. Giri*

Computational Methods in Statistics and Econometrics, *Hisashi Tanizaki*

Applied Sequential Methodologies: Real-World Examples with Data Analysis, *edited by Nitis Mukhopadhyay, Sujay Datta, and Saibal Chattopadhyay*

Handbook of Beta Distribution and Its Applications, *edited by Richard Guarino and Saralees Nadarajah*

Item Response Theory: Parameter Estimation Techniques, Second Edition, *edited by Frank B. Baker and Seock-Ho Kim*

Statistical Methods in Computer Security, *William W. S. Chen*

Elementary Statistical Quality Control, Second Edition, *John T. Burr*

Data Analysis of Asymmetric Structures, *edited by Takayuki Saito and Hiroshi Yadohisa*

Mathematical Statistics with Applications, *Asha Seth Kapadia, Wenyaw Chan, and Lemuel Moyé*

Advances on Models, Characterizations and Applications, *N. Balakrishnan, I. G. Bairamov, and O. L. Gebizlioglu*

Survey Sampling: Theory and Methods, Second Edition, *Arijit Chaudhuri and Horst Stenger*

Statistical Design of Experiments with Engineering Applications, *Kamel Rekab and Muzaffar Shaikh*

Advances on Models, Characterizations and Applications

N. Balakrishnan
McMaster University
Hamilton, ON, Canada

I.G. Bairamov
Izmir University of Economics
Balcova, Izmir, Turkey

O.L. Gebizlioglu
Ankara University
Tandogan, Ankara, Turkey

CRC Press
Taylor & Francis Group
Boca Raton London New York

CRC Press is an imprint of the
Taylor & Francis Group, an **informa** business
A CHAPMAN & HALL BOOK

CRC Press
Taylor & Francis Group
6000 Broken Sound Parkway NW, Suite 300
Boca Raton, FL 33487-2742

First issued in paperback 2019

© 2005 by Taylor & Francis Group, LLC
CRC Press is an imprint of Taylor & Francis Group, an Informa business

No claim to original U.S. Government works

ISBN-13: 978-0-8247-4022-1 (hbk)
ISBN-13: 978-0-367-39270-3 (pbk)
Library of Congress Card Number 2005041422

Library of Congress Cataloging-in-Publication Data

Advances on models, characterizations, and applications / [edited by] N. Balakrishnan, I.G. Bairamov, and O.L. Gebizlioglu.
    p. cm.
Includes bibliographical references and index.
ISBN 0-8247-4022-X
    1. Distribution (Probability theory) 2. Statistical hypothesis testing--Mathematical models. I. Balakrishnan, N. II. Bairamov, I. G. III. Gebizlioglu, O. L.

QA273.6.A39 2005
519.2'4--dc22

2005041422

Visit the Taylor & Francis Web site at
http://www.taylorandfrancis.com

and the CRC Press Web site at
http://www.crcpress.com

Preface

Statistical distributions and models play a vital role in many different applied fields in science, engineering, humanities, health and social sciences. Data obtained from real-life situations are often modeled by appropriate statistical distributions and then inferential procedures are developed under the assumption of that particular distribution. For this reason, it becomes very important that properties of statistical distributions are studied so that they can be utilized to develop optimal inferential methods for analyzing the data under the considered statistical model, and also to check the validity of that model assumption for the data at hand.

Many authoritative and encyclopedic volumes on statistical distribution theory exist in the literature. This list includes:

- Johnson, Kotz and Kemp (1992), describing discrete univariate distributions
- Stuart and Ord (1993), discussing general distribution theory
- Johnson, Kotz and Balakrishnan (1994, 1995), describing continuous univariate distributions

- Johnson, Kotz and Balakrishnan (1997), describing discrete multivariate distributions
- Wimmer and Altmann (1999), presenting a thesaurus on discrete univariate distributions
- Evans, Peacock and Hastings (2000), describing discrete and continuous distributions
- Kotz, Balakrishnan and Johnson (2000), discussing continuous multivariate distributions
- Balakrishnan and Nevzorov (2003), providing an introductory exposition to distribution theory
- Zelterman (2004), discussing discrete distributions and their applications in health sciences

All these books/volumes provide ample evidence of the importance of this area of research.

In this volume, we present 14 chapters written by internationally renowned experts. These chapters discuss characterizations and other important properties of several statistical distributions and models, inferential procedures for these distributions and models, and finally some applications to real-life problems. Each chapter has been written in a self-contained expository manner with a comprehensive list of pertinent references. These chapters are based on some selected papers that were presented at the *International Conference on Advances on Models, Characterizations and Applications* that was held in Antalya, Turkey, in December 2001.

It is our sincere hope that readers of this volume will get up-to-date information on some recent developments on characterizations and other important properties of several distributions, on some inferential issues relating to these models, and finally on some applications of these models to real-life problems.

We thank all the authors for presenting their work in this volume and also for their support and cooperation. We gratefully acknowledge the help of the referees. Our final special thanks go to Ms. Maria Allegra and Mr. Kevin Sequeira of Marcel Dekker for their support and encouragement, Ms. Preethi Cholmondeley of CRC Press – Taylor & Francis for helping us with the production of the volume, and Mrs. Debbie Iscoe for her fine work in typesetting the entire volume.

N. Balakrishnan
HAMILTON, CANADA

I. Bairamov
İZMİR, TURKEY

O. Gebizlioglu
ANKARA, TURKEY

October 2004

Contents

Contributors

M. Ahsanullah
Department of Management
  Sciences
Rider University
Lawrenceville, NJ 08648, U.S.A.
ahsan@rider.edu

Esra Akdeniz
Department of Statistics, Faculty
  of Arts and Sciences
University of Cukurova
01330, Adana, Turkey
esraakdeniz1@yahoo.com

Fikri Akdeniz
Department of Statistics, Faculty
  of Arts and Sciences
University of Cukurova
01330, Adana, Turkey
akdeniz@mail.cu.edu.tr

Barry C. Arnold
Department of Statistics,
  University of California
Riverside
CA 92521, U.S.A.
barry.arnold@ucr.edu

G. Arslan
Department of Statistics and
  Computer Sciences
Başkent, University
06530 Ankara, Turkey
guvenca@baskent.edu.tr

Majid Asadi
Department of Statistics
University of Isfahan
Isfahan, 81744, Iran
M.Asadi@sci.ui.ac.ir

I. G. Bairamov
Department of Mathematics
İzmir University of
Economics, 35330, Balcova, İzmir,
  Turkey
bayramov@science.ankara.
  edu.tr

N. Balakrishnan
Department of Mathematics and
  Statistics
McMaster University
Hamilton, Ontario L8S 4K1,
  Canada
bala@univmail.cis.
  mcmaster.ca

Konstancja Bobecka
Wydział Matematyki i Nauk
  Informacyjnych
Politechnika Warszawska
Plac Politechniki 1, 00-661
  Warszawa, Poland
bobecka@mini.pw.edu.pl

Ch. A. Charalambides
Department of Mathematics
University of Athens
GR-15784 Athens, Greece
ccharal@math.uoa.gr

Erhard Cramer
Institute of Statistics
RWTH Aachen University
52056 Aachen, Germany
erhard.cramer@
  rwth-aachen.de

Carles M. Cuadras
Department of Statistics
University of Barcelona
08028 Barcelona, Spain
carlesm@porthos.bio.ub.es

O. Gebizlioglu
Department of Statistics
Ankara University
06100, Tandogan, Ankara, Turkey
gebizli@science.ankara.
  edu.tr

Udo Kamps
Institute of Statistics
RWTH Aachen University
52056 Aachen, Germany
udo.kamps@rwth-aachen.de

S. V. Malov
Department of Mathematics and
  Mechanics
St. Petersburg State University
198504, Bibliotechnaya Sq. 2,
  St. Petersburg, Russia
Sergey.Malov@pobox.spbu.ru

N. Papadatos
Department of Mathematics
University of Athens
GR-15784 Athens, Greece
npapada@cc.uoa.gr

Ludger Rüschendorf
Department of Mathematical
  Stochastics
University of Freiburg
Eckerstr. 1, D–79104 Freiburg,
  Germany
ruschen@stochastik.uni-
  freiburg.de

Mehdi Razzaghi
Department of Mathematics,
  Computer Science and
Statistics, Bloomsburg University
Bloomsburg, PA 17815, U.S.A.
razzaghi@bloomu.edu

Masaaki Sibuya
Business Administration,
   Takachiho University
Ohmiya, Suginami-ku
168-8508 Tokyo, Japan
sibuyam@takachiho.ac.jp

Jacek Wesołowski
Wydział Matematyki i Nauk
   Informacyjnych
Politechnika Warszawska
Plac Politechniki 1, 00-661
   Warszawa, Poland
wesolo@mini.pw.edu.pl

Chapter 1

The Shapes of the Probability Density, Hazard, and Reverse Hazard Functions

MASAAKI SIBUYA
Takachiho University, Tokyo

CONTENTS

ABSTRACT

The shapes of a probability density function, its hazard function, and its reverse hazard function restrict each other. Here, "shape" means six types of the graphs of a continuous univariate function: increasing (i, for short), decreasing (d), unimodal (id), anti-unimodal (di), increasing–decreasing–increasing (idi), and decreasing–increasing–decreasing (did). It is proved that among 216 combinations of the shapes of the three functions, 44 cases are possible.

This result is a nonparametric characterization of a triplet, the probability density, hazard and reverse hazard functions, by their shapes.

KEYWORDS AND PHRASES: Characterization, dual failure function, failure rate, hazard rate, logistic distribution, Pareto distribution, reversed hazard rate, Weibull distribution

1.1 INTRODUCTION

The reverse hazard function (r.h.f.), or the reversed hazard rate function, is the ratio of a probability density function (p.d.f.) to its distribution function (d.f.); it is used for the retrospective analysis of survival data. It was introduced by Keilson and Sumita (1982) and called the dual failure function. It has the properties dual to the hazard function (h.f.); see Shaked and Shanthikumar (1994).

Lagakos et al. (1988) used the r.h.f. for a retrospective analysis of epidemiological data on individuals in a group, who are identified by some event and the random time of an initiating event is recorded. Kalbfleisch and Lawless (1989) studied the same kind of data and suggested using the r.h.f. Block et al. (1998) proved, among others, that if a r.h.f. is increasing, its h.f. is also increasing, and its distribution range is limited to $(-\infty, \omega)$, $\omega < \infty$. Hence, the lifetime never has an increasing r.h.f. Based on this fact, they cautioned misuses of the r.h.f.

The shapes of a triplet of a probability density, its hazard, and its reverse hazard functions restrict each other. A "shape" means here a class of piecewise monotone continuous positive functions. In this paper six shapes, on at most three consecutive common intervals (see Subsection 1.3.3), are examined: increasing (i, for short), decreasing (d), unimodal (id), antiunimodal (di), increasing–decreasing–increasing (idi), and decreasing–increasing–decreasing (did). The main result is that among $6 \times 6 \times 6 = 216$ combinations of the shapes of the triplet, 44 cases in Tables 1.3.3 and 1.3.4, Section 1.3, are possible. This result is a nonparametric characterization of a triplet, the probability density, hazard and reverse hazard functions, by their shapes, and can be used as a reference for the modeling based on hazard and reverse hazard functions.

Figure 1.1.1 shows two examples of triplets with different combinations of shapes.

This paper extends a previous one on the relation between the shapes of a probability density and its hazard functions (Sibuya, 1996) and completely solves the problem raised by Block et al. (1998).

The shapes of a h.f. were studied by Aalen and Gjessing (2001) from a different point of view. Monotone h.f. and r.h.f.

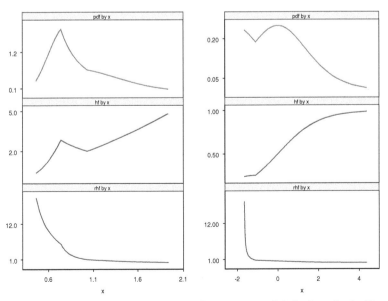

Figure 1.1.1 Examples A and B (from top: p.d.f., h.f. and r.h.f.).

were discussed in Sengupta and Nanda (1999) and Chandra and Roy (2001). An ordering of lifetime distributions by the increasing ratio of a pair of h.f.'s, or a pair of r.h.f.'s, was discussed by Sibuya and Suzuki (2001).

1.2   MONOTONE SHAPES

The h.f. is used mainly in the lifetime analysis and for positive random variables. Here, random variables are not restricted to be positive. The distribution limits of a d.f. F are defined by $\alpha := \inf\{x; F(x) > 0\}$, and $\omega := \sup\{x; F(x) < 1\}$, $-\infty \le \alpha < \omega \le \infty$. It is assumed that the p.d.f. satisfies $f(x) > 0$, $\alpha < x < \omega$. The shapes are invariant with respect to the piecewise change of location and scale. Hence, whether the limits are finite or infinite is the concern. The symbols α and ω are overused within tables to mean finite limits.

1.2.1   Definitions and duality

Let $\bar{F}(x) = 1 - F(x)$ be a survival function (s.f.); the h.f. is defined by

$$h(x) := \frac{d}{dx}(-\log(\bar{F}(x))) = f(x)/\bar{F}(x) \ge 0, \quad \alpha < x < \omega.$$

$$(1.2.1)$$

Conversely, the cumulative h.f. determines its d.f.:

$$\bar{F}(x) = \exp(-H(x)), \qquad H(x) = \int_{-\infty}^{x} h(t)\,dt.$$

$H(x)$ is increasing, $H(\alpha) = 0$, and $H(\omega) = \infty$.
    The r.h.f. is a dual of h.f.:

$$\tilde{h}(x) = \frac{d}{dx}(\log F(x)) = f(x)/F(x) \ge 0, \quad \alpha < x < \omega,$$

$$(1.2.2)$$

$$F(x) = \exp(-\tilde{H}(x)) \qquad \tilde{H}(x) = \int_{x}^{\infty} \tilde{h}(t)\,dt.$$

$\tilde{H}$ is decreasing, $\tilde{H}(\alpha) = \infty$, and $\tilde{H}(\omega) = 0$. Note that if $\alpha > -\infty$ it is possible that $H(\alpha+0) > 0$, and if $\omega < \infty$ it is possible that $\tilde{H}(\omega - 0) > 0$. The cumulative h.f. and r.h.f. are directly related:

$$H(x) = \Omega(\tilde{H}(x)) \quad \text{and} \quad \tilde{H}(x) = \Omega(H(x)), \text{where}$$
$$\Omega(t) = -\log(1 - \exp(-t)), \quad 0 < t < \infty.$$

PROPOSITION 1.2.1 (reflection or duality) *If the p.d.f., d.f., s.f., h.f., r.h.f., cumulative h.f., and cumulative r.h.f. of a r.v. X are $f(x)$, $F(x)$, $\bar{F}(x)$, $h(x)$, $\tilde{h}(x)$, $H(x)$, and $\tilde{H}(x)$, respectively, those of the negated $-X$ are $f(-x)$, $\bar{F}(-x)$, $F(-x)$, $\tilde{h}(-x)$, $h(-x)$, $\tilde{H}(-x)$, and $H(-x)$, respectively. (Note the change of order.)*

This simple fact will be repeatedly used in this paper.

1.2.2   Monotone h.f. and r.h.f.

Before discussing monotone shapes, note that, for a constant $\lambda > 0$,

If $h(x) = \lambda$, $\bar{F}(x) = e^{-\lambda x}/\bar{F}(a)$ and $\tilde{h}(x) \downarrow$, $x > a$.
If $\tilde{h}(x) = \lambda$, $F(x) = e^{\lambda x}/F(b)$ and $h(x) \uparrow$, $x < b$.
If $f(x) = \lambda$, $F(x) = F(a) + \lambda(x - a)$, $\bar{F}(x) = \bar{F}(b) + \lambda(b - x)$
   and $h(x) \uparrow$, $\tilde{h}(x) \downarrow$, $a < x < b$.

Throughout the paper the terms *increasing* and *decreasing* are used in a weak sense.
   From the definitions and Proposition 1.2.1,

$h \uparrow (\downarrow) \Leftrightarrow H$: convex(concave),   and
$\tilde{h} \downarrow (\uparrow) \Leftrightarrow \tilde{H}$: convex(concave).

H : concave $\Rightarrow \tilde{H}$: convex,   and
$\tilde{H}$: concave $\Rightarrow H$: convex.

PROPOSITION 1.2.2

(i)   *If a h.f. is decreasing, its p.d.f. is decreasing, which implies its r.h.f. is decreasing.*

(ii)   *Similarly, if a r.h.f. is increasing, its p.d.f. is increasing, which implies its h.f. is increasing.*

PROOF. From the definition of h and $\tilde{h}$, a pair of fundamental inequalities are obtained:

$$(\log f)' = (\log h)' - h \le (\log h)',$$
$$(\log f)' = (\log \tilde{h})' + \tilde{h} \ge (\log \tilde{h})'. \tag{1.2.3}$$

The first one implies $f \uparrow \Rightarrow h \uparrow$ and $h \downarrow \Rightarrow f \downarrow$, and the second implies $f \downarrow \Rightarrow \tilde{h} \downarrow$ and $\tilde{h} \uparrow \Rightarrow f \uparrow$. Combine these to show the proposition. ∎

REMARK 1.2.1

1. The inequalities hold at any $x \in (\alpha, \omega)$; that is, the proposition states local properties.
2. The second half (ii) of the proposition is dual to (i).

PROPOSITION 1.2.3 (restriction of distribution range) *Let t be any number such that $\alpha < t < \omega$.*

(i) *If a h.f. is decreasing in (α, t), $\alpha > -\infty$. If a h.f. is decreasing in (t, ω), $\omega = \infty$.*
(ii) *If a r.h.f. is increasing in (α, t), $\alpha = -\infty$. If a r.h.f. is increasing in (t, ω), $\omega < \infty$.*

PROOF. If a p.d.f. is decreasing in the lower tail, *a fortiori* if a h.f. is decreasing in the lower tail, its lower distribution limit is finite, because that $f(t) \downarrow$ on $(-\infty, t)$ is impossible. Since $H(\omega) = \infty$, a decreasing h.f. cannot end at a finite point. The second half (ii) is dual to (i). ∎

REMARK 1.2.2 The facts of (i) can be confirmed by observing the cumulative h.f., H. If h is decreasing in (α, t), H is concave and increasing, hence $\alpha > -\infty$. If H is concave in (t, ω), $H(\omega)$ cannot be infinite unless $\omega = \infty$.

1.2.3   Truncation

PROPOSITION 1.2.4 (truncation)

(i) *If a p.d.f. is left-truncated at a, $\alpha < a < \omega$, its h.f. does not change in (a, ω). Further, if the original r.h.f. is decreasing, the new r.h.f. is also decreasing in (a, ω).*

(ii) If a p.d.f. is right-truncated at b, $\alpha < b < \omega$, its r.h.f. does not change in (α, b). Further, if the original h.f. is increasing, the new h.f. is also increasing in (α, b).

PROOF. For the original $f, F, \bar{F}, h,$ and $\tilde{h}$, let the left-truncated be denoted with an asterisk:

$$f^*(x) = f(x)/\bar{F}(a), \qquad \bar{F}^*(x) = \bar{F}(x)/\bar{F}(a),$$
$$F^*(x) = (F(x) - F(a))/(1 - F(a)), \quad x > a.$$

Hence,

$$h^*(x) = h(x), \qquad \tilde{h}^*(x) = \frac{1 - F(a)}{1 - F(a)/F(x)}\tilde{h}(x), \quad x > a,$$

and the latter is decreasing. The second half (ii) is dual to (i). ∎

1.3 NECESSITY OF RESTRICTIONS

1.3.1 The shapes of a doublet of p.d.f. and h.f. (or r.h.f.)

Before discussing the shapes of a triplet, combinations of the shapes of a doublet of p.d.f. and h.f., or a doublet of p.d.f. and r.h.f., are examined. Propositions 1.2.2 and 1.2.3 restrict possible combinations and the range.

Impossible combinations of the shapes of p.d.f. and h.f. are shown in Table 1.3.1. For other combinations, the necessary restrictions on the distribution range are shown. Similarly, impossible combinations of the shapes of p.d.f. and r.h.f., as well as restrictions on the range, are shown in Table 1.3.2. Table 1.3.1 is a modification of a previous table for the lifetime distributions (Sibuya, 1996).

Table 1.3.1 is constructed along the following rules because of Theorems 1.3.3 and 1.3.4.:

(i) If h is decreasing in some interval (symbol d) and f is increasing (symbol i) in this interval, the doublet is impossible (symbol †).

(ii) If h is decreasing in the upper tail (f is also decreasing), $\omega = \infty$ (symbol +).

TABLE 1.3.1   Possible combinations of the shapes of h.f. and its p.d.f.

$h\backslash f$	i	d	id	di	idi	did
i	ω	α	$\cdot$	$\alpha\omega$	ω	α
d	$\dagger$	$\alpha+$	$\dagger$	$\dagger$	$\dagger$	$\dagger$
id	$\dagger$	$\alpha+$	$+$	$\dagger$	$\dagger$	$\alpha+$
di	$\dagger$	α	$\dagger$	$\alpha\omega$	$\dagger$	α
idi	$\dagger$	α	$\boxed{\begin{matrix}\text{idd}\\ \cdot\end{matrix}}$	$\boxed{\begin{matrix}\text{ddi}\\ \alpha\omega\end{matrix}}$	ω	$\dagger*$
did	$\dagger$	$\alpha+$	$\dagger$	$\dagger$	$\dagger$	$\alpha+$

$\dagger$: impossible, $*$: see Section 1.3.3, $\cdot$: any, $-$: $\alpha = -\infty$,
$+$: $\omega = \infty$, α: $\alpha > -\infty$, ω: $\omega < \infty$.

(iii)   If f is decreasing in the lower tail, α is finite (symbol α), and if f is increasing in the upper tail, ω is finite (symbol ω).

Take, for example, a row (h: id) of Table 1.3.1. Since $h \downarrow$ in (t, ω) and $f \downarrow$ in (t, ω), the shapes i, di, and idi of f are impossible (symbol $\dagger$), and in the other entries d, id, and did, $\omega = \infty$, the symbol $+$. Further, f is decreasing in the lower tail in the entry f: d and did, $\alpha > -\infty$, the symbol α, independently of the shape of h.

For another example, take the case (h: idi). As the shape of h changes in three consecutive intervals, the corresponding shape (f: id) can be iid or idd. The first, (f: iid), is impossible, since f should be d in the middle part. For the second, (f: idd),

TABLE 1.3.2   Possible combinations of the shapes of r.h.f. and its p.d.f.

$\hbar\backslash f$	i	d	id	di	idi	did
i	$-\omega$	$\dagger$	$\dagger$	$\dagger$	$\dagger$	$\dagger$
d	ω	α	$\cdot$	$\alpha\omega$	ω	α
id	$-\omega$	$\dagger$	$-$	$\dagger$	$-\omega$	$\dagger$
di	ω	$\dagger$	$\dagger$	$\alpha\omega$	ω	$\dagger$
idi	$-\omega$	$\dagger$	$\dagger$	$\dagger$	$-\omega$	$\dagger$
did	ω	$\dagger$	$\boxed{\begin{matrix}\text{iid}\\ \cdot\end{matrix}}$	$\boxed{\begin{matrix}\text{dii}\\ \alpha\omega\end{matrix}}$	$\dagger*$	α

$\dagger$: impossible, $*$: see Section 1.3.3, $\cdot$: any, $-$: $\alpha = -\infty$,
$+$: $\omega = \infty$, α: $\alpha > -\infty$, ω: $\omega < \infty$.

there is no restriction of the range in both lower and upper ranges. Another shape (f: di) can be dii or ddi, and the first is impossible; the second shape (f: ddi) has restrictions in both tails, and the limits are finite.

Table 1.3.2 is constructed similarly along the following rules:

(i)   If $\tilde{h}$ is increasing in some interval, and f is decreasing in this interval, the doublet is impossible.

(ii) If $\tilde{h}$ is increasing in the lower tail, $\alpha = -\infty$ (symbol $-$).

(iii) The conditions on the distribution range based on f are the same as Table 1.3.1.

Table 1.3.2 is also constructed from Table 1.3.1 according to the following rules based on Proposition 1.2.1.

Change $f(x)$ to $f(-x)$ and, accordingly, exchange (f: i) and (f: d), the conditions (f: id, di, idi, and did) are not changed; and change (h: i (or d)) to ($\tilde{h}$: d (or i)).

In the entries, change α(or $\alpha+$) to ω(or $-\omega$), and exchange $+$ and $-$.

1.3.2   The shapes of a triplet

Specify a pair of shapes of h.f. and r.h.f., for each shape restrict possible shapes of the p.d.f. and the range, and the triplet is impossible if the intersection of two possible ranges is empty. If the intersection is not empty, the intersection is a necessary condition of the distribution range. The results are summarized in Tables 1.3.3 and 1.3.4.

TABLE 1.3.3   Possible combinations of the shapes of h.f. and its r.h.f.

$h\backslash\tilde{h}$	i	d	id	di	idi	did
i	(1)	(3)	(4)	(6)	(10)	(12)
d	†	(2)	†	†	†	†
id	†	(5)	(8)	†	†	(14)
di	†	(7)	†	(9)	†	(16)
idi	†	(13)	(15)	(17)	(18)	†*
did	†	(11)	†	†	†	(19)

†: impossible, *: see Section 1.3.3.
(number): possible cases shown in Table 1.3.4.

TABLE 1.3.4   Possible combinations of the shapes of a triplet $(h, \hbar, f)$ and its range.

No.	h	$\hbar$	f					
			i	d	id	di	idi	did
(1)	i	i	$-\omega$	†	†	†	†	†
(2)	d	d	†	$\alpha+$	†	†	†	†
(3)	i	d	ω	α	·	$\alpha\omega$	ω	α
(4)	i	id	$-\omega$	†	—	†	$-\omega$	†
(5)	id	d	†	$\alpha+$	+	†	†	$\alpha+$
(6)	i	di	ω	†	†	$\alpha\omega$	ω	†
(7)	di	d	†	α	†	$\alpha\omega$	†	α
(8)	id	id	†	†	$-+$	†	†	†
(9)	di	di	†	†	†	$\alpha\omega$	†	†
(10)	i	idi	$-\omega$	†	†	†	$-\omega$	†
(11)	did	d	†	$\alpha+$	†	†	†	$\alpha+$
(12)	i	did	ω	†	·	$\alpha\omega$	†*	α
(13)	idi	d	†	α	·	$\alpha\omega$	ω	†*
(14)	id	did	†	†	+	†	†	$\alpha+$
(15)	idi	id	†	†	—	†	$-\omega$	†
(16)	di	did	†	†	†	$\alpha\omega$	†	α
(17)	idi	di	†	†	†	$\alpha\omega$	ω	†
(18)	idi	idi	†	†	†	†	$-\omega$	†
(19)	did	did	†	†	†	†	†	$\alpha+$

†: impossible, *: see Section 1.3.3, ·: any, $-$: $\alpha = -\infty$, $+$: $\omega = \infty$, α: $\alpha > -\infty$, ω: $\omega < \infty$. (44 possible cases)

TABLE 1.3.5   Comparison of Tables 1.3.1 and 1.3.2 (first example).

No. 5	f	i	d	id	di	idi	did	
h	id	†	$\alpha+$	+	†	†	$\alpha+$	: From Table 1.3.1
$\hbar$	d	ω	α	·	$\alpha\omega$	ω	α	: From Table 1.3.2
join		†	$\alpha+$	+	†	†	$\alpha+$	

TABLE 1.3.6   Comparison of Tables 1.3.1 and 1.3.2 (second example).

	f	i	d	id	di	idi	did	
h:	idi	†	α	$\boxed{\begin{array}{c}\text{idd}\\ \cdot\end{array}}$	$\boxed{\begin{array}{c}\text{ddi}\\ \alpha\omega\end{array}}$	ω	†*	: From Table 1.3.1
$\hbar$:	did	ω	†	$\boxed{\begin{array}{c}\text{iid}\\ \cdot\end{array}}$	$\boxed{\begin{array}{c}\text{dii}\\ \alpha\omega\end{array}}$	†*	α	: From Table 1.3.2
join		†	†	†	†	†*	†*	* : See Section 1.3.3

The way to construct Tables 1.3.3 and 1.3.4 from Table 1 is explained by an example, an entry of (h: id, $\hbar$: d). Take the row of (h: id) from Table 1.3.1 and the row of ($\hbar$: d) from Table 1.3.2 and compare them as follows.

For each shape of p.d.f., take the more restrictive range to get the possible shape of the p.d.f. and the necessary condition of the range. The procedure is shown in Table 1.3.5.

Another example is entry(h: idi, $\hbar$: did). In this case (f: id) and (f: di) are possible for both (h: idi) and ($\hbar$: did). However, more precisely, (f: iid) is possible for (h: idi) and (f: iid) for ($\hbar$: did); hence, (f: id) is impossible. By a similar reason, (f: di) is impossible. Hence, (h: idi, $\hbar$: did) is impossible. The procedure is shown in Table 1.3.6.

Results of the pairing are summarized in Tables 1.3.3 and 1.3.4. The column ($\hbar$: i) of Table 1.3.3, combined with the row (1) of Table 1.3.4 (its dual is the row (h: d) of Table 1.3.3 and the row (2) of Table 1.3.4), shows the result by Block et al. (1998): If h is increasing, $\tilde{h}$ should be so, and vice versa, and moreover the range is restricted to $(-\infty, \omega)$. Example distributions of some entries are shown in Table 1.5.1 of Section 1.5. If the distribution is limited to lifetime, the entries with symbol — in Table 1.3.4 are impossible, the number of possible cases is reduced to 31, and 12 cases among them are restricted to $\omega < \infty$.

1.3.3   Four-interval description

In Table 1.3.1 the case (h: idi, f: did) is regarded impossible, and in Table 1.3.2 the case ($\hbar$: did, f: idi) is regarded impossible, because the combinations of middle term are impossible. However, the cases are made possible if four intervals are used to describe the shapes idi or did. The following are extensions of Table 1a and Table 1b. The impossible combinations in

Table 1.3.1 (extension).

	$h \backslash f$	did	
		didd	ddid
idi	iidi	α	†
	idii	†	α

Table 1.3.2 (extension).

	$\hbar \backslash f$	idi	
		iidi	idii
did	didd	ω	†
	ddid	†	ω

the middle term are avoided by lagged patterns with an extra interval.

If the trick is included, the entry (*h*: idi, $\tilde{h}$: did) in Table 1.3.3 is possible: The discussion of the previous subsection on this pattern should be changed. This means, in Table 1.3.4, to add a new line (20). Moreover, two entries, the row (12) column idi and the row (13) column did, are made possible. The extension of Table 1.3.4 is as follows:

Table 1.3.4 (extension).

No.	h	f / $\tilde{h}$	idi		did	
			iidi	idii	didd	ddid
(12)	i	did — didd	ω	†	α	†
		did — ddid	†	ω	†	α
(13)	idi — iidi	d	ω	†	α	†
	idi — idii	d	†	ω	†	α
(20)	iidi	didd	ω	†	α	†
	idii	ddid	†	ω	†	α

These cases are not included in Tables 1.3.1–1.3.4, mainly because they become messy. Talking only on shapes, one could say "there are $44 + 4$ possible cases," disregarding four-interval details. They are possible within lifetime distributions.

1.4 SUFFICIENCY OF RESTRICTIONS

In the previous section, it is shown that the conditions in Tables 1.3.3 and 1.3.4 are necessary for a triplet. In this section, it is shown that for every condition a triplet really exists. That is, given a possible pair of shapes of the h.f. and r.h.f., a set of possible shapes of p.d.f. is constructed explicitly depending on the distribution range.

1.4.1 Cut and paste of a triplet

PROPOSITION 1.4.1 (cut and paste) *Let f_i, $i = 1, 2$, be p.d.f.'s; F_i their d.f.'s; (a_i, b_i) quantiles such that $F_i(a_i) = p_1$ and $F_i(b_i) = p_2$, $0 \le p_1 < p_2 \le 1$. It is possible to construct a new p.d.f. f_2^* with d.f. F_2^* such that $F_2^*(a_2) = p_1$, $F_2^*(b_2) = p_2$, and $f_2^*(x)$ has the similar shape as $f_2(x)$ in (b_2, ω) and in (α, a_2), and the similar shape as $f_1(x)$, in (a_2, b_2).*

PROOF. Cut the p.d.f. $f_1(x)$ in the range $a_1 \le x \le b_1$, rescale and shift it, and replace $f_2(x)$ in the range (a_2, b_2):

$$f_2^*(x) := \frac{b_2 - a_2}{b_1 - a_1} f_1\left(a_2 + \frac{b_2 - a_2}{b_1 - a_1}(x - a_1)\right), \quad a_2 \le x \le b_2.$$

The gaps between $f_2^*(x)$ and $f_2(x)$ at $x = a_2$ and b_2 are adjusted:

$$f_2^*(x) := \begin{cases} \frac{f_2^*(b)}{f_2(b)} f_2\left(b + \frac{f_2(b)}{f_2^*(b)}(x - b)\right), & x > b, \\ \frac{f_2^*(a)}{f_2(a)} f_2\left(a - \frac{f_2(a)}{f_2^*(a)}(a - x)\right), & x < a. \end{cases} \quad (1.4.1)$$

∎

REMARK 1.4.1 The new h.f. and r.h.f. are also similar to those of f_2 in (b_2, ω) and (α, a_2) and to those of f_1 in (a_2, b_2). Hence, in any interval a triplet can be replaced by that part of another triplet, keeping its shapes.

1.4.2 Examples of monotone shapes

In this subsection, cases where all the shapes of the triplet are monotone, i.e., the entries (1), (2), and a part of (3) of Tables 1.3.3 and 1.3.4, are discussed. Further, the discussion is extended to the remaining part of (3).

The first case to be considered is $(h: \mathrm{d}, \tilde{h}: \mathrm{d})$ with the range (α, ∞). Typical distributions of this type are the Pareto,

$$\bar{F}(x) = x^{-\gamma}, \quad h(x) = \gamma/x, \quad \tilde{h}(x) = h(x)/(x^\gamma - 1);$$
$$x > 1, \gamma > 0, \quad (1.4.2)$$

and the Weibull (with the power parameter less than one),

$$\bar{F}(x) = \exp(-x^{\gamma}), \quad h(x) = \gamma/x^{\gamma-1},$$
$$\hbar(x) = h(x)/(\exp(x^{\gamma}) - 1); \quad x > 0, \gamma < 1. \tag{1.4.3}$$

The second case (h: i, $\hbar$:i) with the range $(-\infty, \omega)$ is dual to (h: d, $\hbar$: d), and the negative Pareto and the negative Weibull are typical examples because of Proposition 1.2.1.

Since (h: d, $\hbar$: i) is impossible, the last monotone case is (h: i, $\hbar$: d). A typical distribution of this property is the logistic distribution. Its density is unimodal and the range is $(-\infty, \infty)$,

$$\bar{F}(x) = 1/(1 + \exp(x)), \quad F(x) = 1/(1 - \exp(-x)),$$
$$f(x) = F(x)\bar{F}(x), \quad h(x) = F(x) \quad \text{and} \tag{1.4.4}$$
$$\hbar(x) = \bar{F}(x); \quad -\infty < x < \infty.$$

Because of Proposition 1.2.4, any truncation keeps the shape (h: i, $\hbar$: d), and the following shapes and ranges of p.d.f. can be constructed:

(f: d) on (α, ∞), (α, ω),

(f: i) on $(-\infty, \omega)$, (α, ω),

(f: id) on $(-\infty, \infty)$, $(-\infty, \omega)$, (α, ∞), (α, ω).

The first two, with the previous two of this subsection, show the construction of the triplets of all monotone shapes and are summarized in Table 1.4.1, including the following shapes.

The starting distribution (1.4.4) of the case (h: i, $\hbar$: d) has (f: id), and its range has no restriction as shown above. The other shapes of f are constructed by the "cut and paste" method of Proposition 1.4.1. From the logistic distribution (1.4.4), let

$$f_1(x) = f(x)/\bar{F}(a), \quad 0 \le a < x; \quad \text{and} \quad f_2(x) = f(x)/F(b),$$
$$x < b \le 0.$$

Replace an upper tail of F_1 by the corresponding upper tail of f_2 to obtain (f: di). Further replacing an upper tail of (f: i) (or a lower tail of (f: d) by a suitable part of (f: di), (f: idi), or (f: did)) is constructed. For (f: idi) (or (f: did)) only the upper

TABLE 1.4.1   Monotone triplets (1), (2), (3) in Table 1.3.3.

h	i	d	i	i	i	i	i	i
$\check{h}$	i	d	d	d	d	d	d	d
f	i	d	i	d	id	di	idi	did
range	$-\omega$	$\alpha+$	ω	α	$\cdot$	$\alpha\omega$	ω	α

(or lower) tail restricts the range, and the range is the same as $(f: i)$ (or $(f: d)$).

Thus, sufficiency of the entries (1), (2), and (3) of Table 1.3.4 is proved.

1.4.3   General shapes

The triplets of monotone shapes are summarized in Table 1.4.1. How to paste them to construct a general triplet was shown in the previous subsection, by studying the case of $(h: i, \check{h}: d)$ for $(f: di, idi, did)$. This cut and paste method can be used for general shapes.

Take, for example, the case of $(h: i, \check{h}: di)$, the entry (6), for $(f: i, di, idi)$ with the necessary range $(\omega, \alpha\omega, \omega)$, respectively. The distribution range is divided into two or three parts corresponding to the shapes of $\check{h}$ or f. Corresponding to $(f: idi)$, $\check{h}$ is divided into $(\check{h}: iid)$ or $(\check{h}: iid)$, but the former is impossible. The restrictions shown in the table are relevant in both ends and are consistent with those at the bottom. Now, for each of $(f: i, di, idi)$, two or three intervals of Table 1.4.1 are pasted to form the triplets of the given shapes. This confirms that the row of $(h: i, \check{h}: di)$ of Table 1.3.4 can be constructed. The method of construction is illustrated in Table 1.4.2.

PROPOSITION 1.4.2 *All the "not impossible" cases in Table 1.3.4 are actually possible.*

TABLE 1.4.2   Construction example, $(h: i, \check{h}: di)$.

interval	h	$\check{h}$	f		f		f	
lower	i	d	i	ω	d	α	i	ω
middle	i	d					d	
upper	i	i	i	$-\omega$	i	$-\omega$	i	$-\omega$
pasted	i	di	i	ω	di	$\alpha\omega$	idi	ω

PROOF. The conditions on the shapes of h and $\hbar$ are divided into at most three intervals. In each interval the combination of h and $\hbar$ is $(i,i), (d,d)$, or (i,d) and the possible shapes of f are listed in Table 1.4.1. Since the conditions of Table 1.4.1 are consistent with those in Table 1.3.3, the set of new pasted p.d.f.'s is the same as in Table 1.3.4. ∎

1.5 ADDITIONAL NOTES

The logic and technique in this paper can be automated and applied to more complex shapes: bimodal, anti-bimodal, and so on.

To determine the shapes of the triplet of a given distribution is beyond the scope of this paper, but shapes of the triplet of some "textbook distributions" are listed in Table 1.5.1. If the shapes of a triplet are limited to four, i, d, id, and di, there are $4 \times 4 \times 4 = 64$ combinations, and 16 cases are possible among them and 10 cases are covered by Table 1.5.1 and the dual distributions (the negation of random variables) of those in Table 1.5.1.

A tool for preparing Table 1.5.1 is the following fact.

TABLE 1.5.1 Some example distributions.

	h	$\hbar$	f	range	distribution
(2)	d	d	d	$(0, \infty)$	gamma (gamma p. < 1), Weibull (power p. < 1), beta2 $(1, \beta)$ = Pareto
(3)	i	d	id	$(-\infty, \infty)$	logistic, normal, Gumbel (double exponential), Laplace (bilateral exponential)
(3)	i	d	id	$(0, \infty)$	gamma (gamma p. > 1), Weibull (power p. > 1)
(5)	id	d	id	$(0, \infty)$	lognormal, Fréchet, beta2 $(\alpha > 1, \beta = 1)$
(7)	di	d	d	$(0, 1)$	beta $(\alpha < 1, \beta = 1)$
(8)	id	id	id	$(-\infty, \infty)$	Cauchy
(9)	di	di	di	$(0, 1)$	beta $(\alpha < 1, \beta < 1)$

TABLE 1.5.2   Transformation of random
variables and the shapes of the doublet.

ψ	X	Y
↑ convex	(h: i, $\hbar$: i)	(h: i, $\hbar$: i)
↑ concave	(h: d, $\hbar$: d)	(h: d, $\hbar$: d)
↓ convex	(h: i, $\hbar$: i)	(h: d, $\hbar$: d)
↓ concave	(h: d, $\hbar$: d)	(h: i, $\hbar$: i)

PROPOSITION 1.5.1 *If a p.d.f. is logconcave on* (α, β), $-\infty \leq \alpha < \beta \leq \infty$, *its h.f. is increasing and r.h.f. is decreasing. If a p.d.f. is logconvex on* $(-\infty, \omega)$, $\omega < \infty$, *its h.f. and r.h.f. are both increasing, and if a p.d.f. is logconvex on* (α, ∞), $\alpha > -\infty$, *its h.f. and r.h.f. are both decreasing.*

Another tool for preparing Table 1.5.1 is the transformation of distributions, including Proposition 1.2.1 as a special case.

PROPOSITION 1.5.2 *Let a r.v. X be a transformation of another r.v. $Y : X = \psi(Y)$. Depending on ψ, the shapes of the h.f. and r.h.f. of Y are determined by those of X, as follows.*

For example, distributions of the shape (h: d, $\hbar$: d) or (h: i, $\hbar$: i) are constructed by transforms starting from one of the shapes. See Table 1.5.2, which illustrates the proposition.

REFERENCES

Aalen, O. O. and Gjessing, H. K. (2001). Understanding the shape of the hazard rate: A process point of view (with discussions). *Statistical Science*, **16**, 1–22.

Block, H. W., Savits, T. H. and Singh, H. (1998). The reversed hazard rate function. *Probability in the Engineering and Informational Sciences*, **12**, 69–90.

Chandra, N. K. and Roy, D. (2001). Some results on reversed hazard rate. *Probability in the Engineering and Information Sciences*, **15**, 95–102.

Kalbfleisch, J. D. and Lawless, J. F. (1989). Inference based on retrospective ascertainment: An analysis of data on transfusion-related

AIDS. *Journal of the American Statistical Association*, **84–406**, 360–372.

Keilson, J. and Sumita, U. (1982). Uniform stochastic ordering and related inequalities. *Canadian Journal of Statistics*, **10**, 181–198.

Lagakos, S. W., Barraj, L. M. and Gruttola, V. De. (1988). Nonparametric analysis of truncated survival data, with application to AIDS. *Biometrika*, **75**, 515–523.

Sengupta, D. and Nanda, A. K. (1999). Log-concave and concave distributions in reliability. *Naval Research Logistics*, **46**, 419–433.

Shaked, M. and Shanthikumar, J. G. (1994). *Stochastic Orders and Their Applications*. Academic Press, Boston.

Sibuya, M. (1996). The shapes of a probability density function and its hazard function. In *Lifetime Data: Models in Reliability and Survival Analysis* (Eds., N. P. Jewell et al.), pp. 307–314, Kluwer Academic, Boston.

Sibuya, M. and Suzuki, K. (2001). Optimal threshold for the k-out-of-n monitor with dual failure modes. *Annals of the Institute of Statistical Mathematics*, **53**, 189–202.

Chapter 2

Stochastic Ordering of Risks, Influence of Dependence, and A.S. Constructions

LUDGER RÜSCHENDORF
Department of Mathematical Stochastics, University of Freiburg, Freiburg, Germany

CONTENTS

ABSTRACT

In this paper we review and extend some key results on the stochastic ordering of risks and on bounding the influence of stochastic dependence on risk functionals. The first part of the paper is concerned with a.s. constructions of random vectors and with diffusion kernel type comparisons which are of importance for various comparison results. In the second part we consider generalizations of the classical Fréchet-bounds, in particular for the distribution of sums and maxima and for more general monotonic functionals of the risk vector. In the final part we discuss three important orderings of risks which arise from Δ-monotone, supermodular, and directionally convex functions. We give some new criteria for these orderings. For the basic results we also take care to give references to "original sources" of these results.

KEYWORDS AND PHRASES: Ordering of risk, supermodular, directionally convex, comonotone, Fréchet-bounds

2.1 INTRODUCTION

It has been recognized in recent years that the methods and tools of stochastic ordering and construction of probabilities with given marginals are of essential relevance for the problem of modeling multivariate portfolios and bounding functions of dependent risks like the value at risk, the expected excess of loss, and other financial derivatives and risk measures. Even if many results on stochastic ordering and dependent risks have been developed in early years, a new impetus on reconsidering this field came recently from financial modeling and risk management, and many papers in economics and insurance journals are devoted to this subject [see, e.g., the recent article of Embrechts, Höing, and Juri (2003) and the references therein]. Stochastic ordering and marginal modeling have a long history and several books and conference proceedings on this subject have appeared, in particular, proceedings of conferences on marginal modeling and stochastic ordering [Dall'Aglio (1972), Mosler and Scarsini (1991a), Rüschendorf, Schweizer, and Taylor, (1996), Beneš and Štěpán (1997), and Cuadras, Fortiana, and Rodriguez-Lallena (2002)] as well as the comprehensive volumes of Stoyan (1977), Marshall and Olkin (1979), Tong (1980), Mosler (1982), Shaked and Shantikumar (1994), Joe (1997), Nelsen (1999), and Müller and Stoyan (2002).

    The main purpose of this paper is to point out and partially extend some of the orderings and results on orderings which seem to be of particular importance for bounding risks and the influence of dependence on functionals. For several of the key results, we also want to give some of the early and original references. The field of stochastic orders is very diversified, but some of the recent work and results have already been stated and established in early papers on stochastic ordering.

We essentially restrict to "integral orders $\prec_{\mathcal{F}}$" on the probability measures induced by some function class $\mathcal{F}$ and defined by

$$P \prec_{\mathcal{F}} Q \quad \text{if} \quad \int f \, dP \le \int f \, dQ \quad \text{for all} \quad f \in \mathcal{F},$$
$$(2.1.1)$$

such that f is integrable w.r.t. P and Q. Some natural questions for the analysis of a stochastic order $\prec$ are to find simple and maximal generators $\mathcal{F}$ of $\prec$ such that $\prec$ and $\prec_{\mathcal{F}}$ are equivalent orderings (or to find at least large classes $\mathcal{F}$ such that $P \prec Q$ implies $P \prec_{\mathcal{F}} Q$). This aspect is discussed in most of the books mentioned above. Additional particular references on the subject of integral stochastic orders are Rüschendorf (1979), Reuter and Riedrich (1981), Mosler and Scarsini (1991b), Marshall (1991), Müller (1997), and Denuit and Müller (2001).

The plan of this paper is to discuss at first a.s. construction of random vectors which lie at the core of several ordering results. Related are kernel representation results which give "pointwise" characterization of stochastic orders by diffusion kernels. Each ordering generates a notion of positive resp. negative dependence by comparing a probability measure

$$P \in M(P_1, \dots, P_n) \qquad (2.1.2)$$

—the class of all probability measures with marginals $P_1, \dots, P_n$—to the product $\otimes_{i=1}^n P_i$ of its marginals. If $\otimes_{i=1}^n P_i \prec P$, then we speak of positive dependence of P; if $P \prec \otimes_{i=1}^n P_i$, then we speak of negative dependence. Related is the problem to describe the maximal influence of dependence on a function f (or class of functionals $\mathcal{F}$),

$$M(f) = \sup \left\{ \int f \, dP; \ P \in M(P_1, \dots, P_n) \right\}$$
$$(2.1.3)$$

$$\text{resp.} \quad m(f) = \inf \left\{ \int f \, dP; \ P \in M(P_1, \dots, P_n) \right\}$$

which we call the problem of *(generalized) Fréchet-bounds.*
One of the most prominent results in stochastic ordering comprises the classical Fréchet-bounds due to Hoeffding (1940)
and Fréchet (1951):

a) For an n-dimensional df F holds: $F \in \mathcal{F}(F_1, \ldots, F_n)$—
    the Fréchet class of n-dimensional df's with marginals
    $F_1, \ldots, F_n$—if and only if

$$F_- \leq F \leq F_+, \qquad (2.1.4)$$

where $F_+(x) := \min_{1 \leq i \leq n}\{F_i(x_i)\}$ and $F_-(x) :=$
$\max\{0, \sum_{i=1}^{n} F_i(x_i) - (n-1)\}$ are the upper and lower
Fréchet-bounds.

b) Moreover, $F_+ \in \mathcal{F}_n$ is an n-dimensional df, while $F_- \in$
    $\mathcal{F}_n$ if and only if $n = 2$ or for $n > 2$

$$\text{either } \sum_{i=1}^{n} F_i(x_i) \leq 1 \quad \text{for all } x \text{ with}$$
$$F_j(x_j) < 1, \ \forall j$$

$$\qquad (2.1.5)$$

$$\text{or} \qquad \sum_{i=1}^{n} F_i(x_i) \geq n - 1 \quad \text{for all } x \text{ with}$$
$$F_j(x_j) > 0, \ \forall j.$$

The important characterization in (b) of the cases where $F_- \in$
$\mathcal{F}_n$ is due to Dall'Aglio (1972). For a review on various aspects
of Fréchet-bounds, see Rüschendorf (1991b) (in the following
abbreviated by Ru (1991b)). The most important general technique to determine generalized Fréchet-bounds is duality theory. A comprehensive survey of this theory and its applications
is given in Rachev and Ru (1998, Vol. I/II). We will describe
some interesting aspects of the problem of Fréchet-bounds on
the influence of dependence in Chapter 3.

For the application to the comparison of risks it has turned
out [see the interesting recent book of Müller and Stoyan (2002)]
that of particular importance are the classes of supermodular
(quasi-monotone), directionally convex, and Δ-monotone func-
tions $\mathcal{F}^{sm}$, $\mathcal{F}^{dcx}$, $\mathcal{F}^{\Delta}$ together with the induced orderings and
some variants like $\mathcal{F}^{idcx}$, the increasing directionally convex
functions. In Section 4 we describe and extend some of the
basic comparison criteria for these functions.

We use some standard notation throughout. $X \sim P$ means
that the random variable X has distribution P. We write for

some ordering $\prec$, $X \prec Y$ synonymously for $P \prec Q$ or $F \prec G$, where F, G are the df's of X, Y and P, Q are the distributions. $\leq_{st}$ denotes the usual stochastic order w.r.t. nondecreasing functions.

2.2 STOCHASTIC ORDERING AND A.S. CONSTRUCTION OF RANDOM VARIABLES

For the comparison of distributions P, Q w.r.t. some stochastic order is in some cases useful to compare explicit a.s. constructions of rv's X, Y where $X \sim P$ and $Y \sim Q$. A general useful construction for $P, Q \in M^1(\mathbb{R}^n)$, the set of probability measures on $\mathbb{R}^n$ is the following "*standard construction*":

Let $F \in \mathcal{F}_n$ be an n-dimensional df and let $V_1, \ldots, V_n$ be independent rv's uniformly distributed on $[0, 1]$, independent of $X \sim F$. Let $V = (V_1, \ldots, V_n)$ and let $F_{i|1, \ldots, i-1}(x_i|x_1, \ldots, x_{i-1})$ denote the conditional df's of X_i given $X_j = x_j, j \leq i-1$. We define $\tau_F : \mathbb{R}^n \times [0, 1]^n \to \mathbb{R}^n$ by

$$\tau_F(x, \lambda) = (F_1(x_1, \lambda_1), F_{2|1}(x_2, \lambda_2|x_1), \ldots,$$

$$F_{n|1, \ldots, n-1}(x_n, \lambda_n|x_1, \ldots, x_{n-1})) \tag{2.2.1}$$

where $F_{i|1, \ldots, i-1}(x_i|x_1, \ldots, x_{i-1}) = P(X_i < x_i|X_j = x_j, j \leq i-1) + \lambda_i P(X_i = x_i|X_j = x_j, j \leq i-1)$. We define the "inverse" transformation τ_F^{-1} recursively as

$$\tau_F^{-1}(u) = z = (z_1, \ldots, z_n), \tag{2.2.2}$$

with $z_1 = F_1^{-1}(u_1)$, $z_2 = \inf\{y : F_{2|1}(y|z_1) \geq u_2\} = F_{2|1}^{-1}(u_2|z_1), \ldots, z_n = F_{n|1, \ldots, n-1}^{-1}(u_n|z_1, \ldots, z_{n-1})$.

THEOREM 2.2.1 (Regression construction) *Let X be an n-dimensional random vector with df F; then:*

a) $U := \tau_F(X, V)$ *has independent components, uniformly distributed on* $[0, 1]$.

b) $Z = \tau_F^{-1}(V)$ *is a rv with df F; Z is called the "regression construction" of F.*

c) $X = \tau_F^{-1}(\tau_F(X, U))$ *a.s.* $\tag{2.2.3}$

REMARK 2.2.1

1) Part (a) of 2.2.1 is due in the case of absolutely continuous conditional df's to Rosenblatt (1952). (a) and (b) were stated in this form in Ru (1981b). (b) had been given before in an equivalent form in O'Brien (1975), and (c) is from Ru and de Valk (1993). The one-dimensional case was used for a long time for the simulation of rv's.

2) By the recursive definition in 2.2.2, one obtains Z also as a function of $(V_1, \dots, V_n)$ which we denote by τ_F^*

$$Z = \tau_F^*(V) \sim F \qquad (2.2.4)$$

where $\tau_F^*(V) = (h_1(V_1), h_2(V_1, V_2), \dots, h_n(V_1, \dots, V_n))$. In this functional form the construction is called the "standard representation" of F. It gives a construction of a random vector with df F as a function of independent uniforms. The functions h_i represent conditional df's.

3) A "copula" of X (resp. F) is any df C with uniform marginals such that

$$C(F_1, \dots, F_n) = F \qquad (2.2.5)$$

where F_i are the marginal df's. If U is a random vector with $U \sim C$, then

$$(F_1^{-1}(U_1), \dots, F_n^{-1}(U_n)) \sim F. \qquad (2.2.6)$$

U represents some aspects of the dependence structure of F (resp. X). To obtain a copula, one can apply Theorem 2.2.1 in the one-dimensional case and consider $\bar{U} := (\tau_{F_1}(X_1, V_1), \dots, \tau_{F_n}(X_n, V_n))$. Then the df C of $\bar{U}$ is a copula and $X = (F_i^{-1}(\bar{U}_i))$ a.s.

We next give some applications of the standard resp. regression construction.

COROLLARY 2.2.1 (Stochastic ordering) *If* $F, G \in \mathcal{F}_n$ *and* $V = (V_1, \dots, V_n)$ *is an iid uniform sequence, then*

$$\tau_F^{-1}(U) \le \tau_G^{-1}(V) \text{ implies } F \le_{st} G, \qquad (2.2.7)$$

where $\leq_{st}$ denotes the usual stochastic ordering w.r.t. $\mathcal{F}^m$ the class of monotonically nondecreasing functions.

REMARK 2.2.2 Condition (2.2.7) is stated in Ru (1981b). It implies various sufficient conditions for stochastic ordering going back to classical results of Veinott (1965), Kalmykov (1962), and Stoyan (1972) in the context of Markov chains. The regression and standard construction are used essentially in various papers on stochastic ordering. The positive dependence ordering "conditional increasing in sequence CIS" just says that the components h_i of τ_F^* are monotonically nondecreasing. This is used essentially in many papers, e.g., in Müller and Scarsini (2001), to state sufficient conditions for the supermodular ordering of positive dependent sequences (see also Section 2.4). An application of the standard construction to convex ordering analogously to (2.2.7) is given in Shaked and Shantikumar (1994). An alternative application to positive regression dependence ordering of rank statistics as well as to further statistical ordering results is given in Ru (1986).

As a second application we consider the following optimal coupling problem: Determine for some $P, Q \in M^1(\mathbb{R}^n)$ and with $S_n(X) = \sum_{i=1}^n X_i$:

$$\inf\{E|S_n(X) - S_n(Y)|^2 : X \sim P, \ Y \sim Q\} \tag{2.2.8}$$

i.e., the problem is to construct two n-dimensional random vectors X, Y with distributions P, Q such that the sums $S_n(X) = \sum_{i=1}^n X_i, S_n(Y) = \sum_{i=1}^n Y_i$ are as close as possible in L^2-distance. The answer to this problem is:

COROLLARY 2.2.2 (Optimal coupling of sums; Ru (1986)) *For $P, Q \in M^1(\mathbb{R}^n)$ let P_1, P_2 denote the distributions of the sums $\sum X_i, \sum Y_i$, vespectively; then*

$$\inf\left\{ E\left|\sum_{i=1}^n X_i - \sum_{i=1}^n Y_i\right|^2 : Y \sim P, \ Y \sim Q \right\} = \ell_2^2(P_1, P_2)$$

$$\tag{2.2.9}$$

where $\ell_2^2(P_1, P_2) = \int_0^1 (F_1^{-1}(u) - F_2^{-1}(u))^2 \, du$ is the squared minimal ℓ_2-metric.

For the proof one applies the regression construction to the extended random vectors $(\sum_{i=1}^{n} X_i, X)$ resp. $(\sum_{i=1}^{n} Y_i, Y)$ and obtains directly (2.2.9). Of course, similar results hold for the coupling of other functionals, such as $(\sum_{i=1}^{n} X_i, \max_{i \le n} X_i)$ simultaneously to $(\sum_{i=1}^{n} Y_i, \max_{i \le n} Y_i)$, etc.

The standard construction does not in general give a pointwise a.s. construction of random vectors $X \sim P$, $Y \sim Q$ such that $X \le Y$ a.s. if $P \le_{st} Q$. But Strassen's comparison theorem implies the existence of such a.s. representations. This result was extended to closed partial orders $\prec$ on a polish space S. The order $\prec$ on S induces the stochastic order $\prec_{st}$ on $M^1(S)$ the set of probability measures on S for the corresponding class $\mathcal{F}^m$ of monotonically nondecreasing functions w.r.t. $\prec$.

THEOREM 2.2.2 (Strassen's theorem) *Let $\prec$ be a closed partial order on a polish space S and P, Q probability measures on S. Then: $P \prec_{st} Q$ if and only if there exist r v's $X \sim P$, $Y \sim Q$ (on some space $(\Omega, \mathcal{A}, R)$) such that*

$$X \prec Y \quad a.s. \tag{2.2.10}$$

REMARK 2.2.3 (2.2.10) was introduced in Strassen (1965) and extended in various ways in Kamae, Krengel, and O'Brien (1978), Kellerer (1984), and Ramachandran and Ru (1995). A proof by means of Strassen's abstract set representation theorem is given in Ru (1980b, Theorem 1) in the case of the Schur-ordering $\prec_S$ on $\mathbb{R}^n$ defined by:

$$a \prec_S b \text{ if } \sum_{i=1}^{k} a_{(i)} \le \sum_{i=1}^{k} b_{(i)}, \quad i \le k \le n-1$$

$$\text{and } \sum_{i=1}^{n} a_i = \sum_{i=1}^{n} b_i, \tag{2.2.11}$$

where $a_{(1)} \ge \ldots \ge a_{(n)}$ are the components arranged in decreasing order. The monotone functions w.r.t. $\prec_S$ are the Schur-convex functions and (2.2.10) implies:

COROLLARY 2.2.3 (Schur-convex ordering) *Let* $P, Q \in M^1(\mathbb{R}^n)$; *then:*

$$P \prec_S Q \Longleftrightarrow \text{There exist } X \sim P, Y \sim Q \text{ with } X \prec_S Y$$

$$(2.2.12)$$

$$\Longleftrightarrow \text{There exist } X \sim P \text{ and a random doubly}$$

$$\text{stochastic matrix } \Pi, \text{ such that } \Pi X \sim Q.$$

A more general comparison result for integral stochastic orders $\prec_{\mathcal{F}}$ is based on $\prec_{\mathcal{F}}$-diffusions.

DEFINITION 2.2.1 ($\prec_{\mathcal{F}}$-diffusions) Let $\mathcal{F}$ be a class of functions on some space $(E, \mathcal{A})$. A Markov kernel K on E is called an $\mathcal{F}$-diffusion if

$$\epsilon_x \prec_{\mathcal{F}} K(x, .) \quad \text{for all } x \in E. \tag{2.2.13}$$

A $\prec_{\mathcal{F}}$-diffusion kernel K "diffuses" locally in any point x mass w.r.t. $\prec_{\mathcal{F}}$. The composition KP is defined by $KP(A) = \int K(x, A) P(dx)$.

PROPOSITION 2.2.1 *Let K be a $\prec_{\mathcal{F}}$-diffusion; then*

$$P \prec_{\mathcal{F}} KP \quad \text{for all } P \in M^1(E). \tag{2.2.14}$$

PROOF. The proof is obvious from the idea of diffusions. For $f \in \mathcal{F}$ holds

$$\int f \, dKP = \int \left(\int f(y) K(x, dy) \right) dP(x)$$

$$\geq \int f(x) \, dP(x). \quad \blacksquare$$

From a general kernel representation result in Strassen (1965), one can obtain a converse of (2.2.14) and characterize several stochastic integral orders by corresponding $\mathcal{F}$-diffusions. A result of this type was stated in Ru (1980b) for several examples including the stochastic order, the convex, and convex increasing order, which were well established before and are related to famous results of Blackwell, Stein, Sherman, Cartier, Meyer, and Strassen. It also included the class of symmetric convex functions and the class of norm increasing functions. The list of examples was further extended in Mosler and Scarsini (1991b). The proof in Ru (1980b) uses an idea from the theory of balagage [Meyer (1966, Theorem 53)] and Strassen's kernel representation theorem [Strassen (1965, Theorem 3)]. For a general formulation of this result we define

$$h_f(x) = \sup\left\{ \int f\,dP; \ \epsilon_x \prec_{\mathcal{F}} P \right\} \tag{2.2.15}$$

for $f \in C_b(E)$ and assume that $\mathcal{F} \subset C_b(E)$ is some order generating class. Let $\mathcal{F}^\circ$ denote the convex maximal generator of the order dual to $\prec_{\mathcal{F}}$ such that $P \prec_{\mathcal{F}} Q$ is equivalent to $Q \prec_{\mathcal{F}^\circ} P$. Let $\overline{\mathcal{F}^\circ}^{P,Q}$ be the set of all pointwise limits of sequences in $\mathcal{F}^\circ$ which are uniformly integrable w.r.t. P, Q.

THEOREM 2.2.3 (*$\mathcal{F}$-diffusions*) *Let E be a polish space, $\mathcal{F} \subset C_b(E)$ with dual convex cone $\mathcal{F}^\circ$, and P, $Q \in M^1(E)$. Assume that for $f \in C_b(E)$, $h_f \in \overline{\mathcal{F}^\circ}^{P,Q}$. Then $P \prec_{\mathcal{F}} Q$ if and only if there exists a $\mathcal{F}$-diffusion K such that*

$$Q = KP. \tag{2.2.16}$$

PROOF. Define $\Pi_x := \{P \in M^1(E) : \epsilon_x \prec_{\mathcal{F}} P\}$. Then Π_x is convex, weakly closed and $h_f(x) = \sup\{\int f\,dP; \ P \in \Pi_x\}$. For $f \in C_b(E)$ holds $f(x) \le h_f(x)$ and, thus, since $h_f \in \overline{\mathcal{F}^\circ}^{P,Q}$

$$\int f\,dQ \le \int h_f\,dQ \le \int h_f\,dP. \tag{2.2.17}$$

This implies by Strassen's kernel representation theorem [Strassen (1965, Theorem 3)] the existence of a kernel K on E with $Q = KP$ and $K(x, .) \in \Pi_x$ for all $x \in E$; i.e., K is an $\mathcal{F}$-diffusion. ∎

REMARK 2.2.4

a) The first general formulation of the diffusion characterization theorem 2.2.16 is due to Meyer (1966, Theorem 53). It is formulated in the context of integral stochastic orders $\prec_{\mathcal{F}}$ in Müller and Stoyan (2002, Theorem 2.6.1): Suppose that for any $f, g \in \mathcal{R}_{\mathcal{F}}$—the maximal generator of $\prec_{\mathcal{F}}$—holds

$$\max(f, g) \in \mathcal{R}_{\mathcal{F}}, \tag{2.2.18}$$

then the equivalence in (2.2.16) holds true.

b) We consider the following examples of applications of (2.2.16):

1) If $\mathcal{F} = \mathcal{F}^{sym,cx}$ is the set of symmetric convex functions on $\mathbb{R}^n$, then h_f is symmetric and concave, so it lies in the closure of the dual cone $\mathcal{F}^{\circ}$. From (2.2.16) we obtain: $P \prec_{sym,cx} Q$ if and only if

$$\exists X \sim P, \ Y \sim Q \text{ such that } X \prec_S E(Y_{(\ )}|X),$$

$$\tag{2.2.19}$$

where $\prec_S$ is the Schur order, $Y_{(\ )}$ is the ordered vector [see Ru (1981)]. It is interesting that there is a difference to the condition for stochastic Schur-ordering in (2.2.12).

2) If $\mathcal{F}^{\|\ \|}$ is the class of norm increasing functions $f(x) = g(\|x\|)$ in $C_b(\mathbb{R}^n)$, then $\epsilon_x \prec_{\mathcal{F}} P$ iff P has support in $\{y : \|y\| \geq \|x\|\}$. Further, for any $f \in C_b$ holds $h_f(x) = \sup\{\int f\,dP : \epsilon_x \prec_{\mathcal{F}^{\|\ \|}} P\}$ is norm decreasing, $\|x\| \leq \|y\|$ implies $h_f(y) \leq h_f(x)$. Thus, $h_f \in \mathcal{F}^{\circ P,Q}$ and we obtain:

$$P \prec_{\mathcal{F}^{\|\ \|}} Q \text{ iff there exist } X \sim P, \ Y \sim Q$$

$$\tag{2.2.20}$$

such that $\|X\| \leq \|Y\|$ a.s. [see Ru (1980b)].

2.3 FRÉCHET BOUNDS—EXTREMAL RISK

As mentioned in the introduction, Fréchet-bounds deal with the basic problem in risk theory to describe the maximal influence of stochastic dependence on the expectation of a functional $\varphi(x_1, \ldots, x_n)$. Examples of interest are, e.g., convex functionals of the joint position $x_1 + \cdots + x_n$, where x_i are risks with distributions P_i. A typical case is $\varphi(x) = (\sum_{i=1}^{n} x_i - k)^+$, the excess of loss function. A nonconvex function of interest is $\varphi_t(x) = 1_{[t,\infty)}(\sum_{i=1}^n x_i)$ which yields just the inverse of the value at risk functional. Of interest is also the maximal risk of the components $\max_{i\le n} x_i$ and variants hereof. For a detailed introduction to these kinds of questions, we refer to Embrechts et al. (2003). An extension of the classical Fréchet-bounds in (2.1.4) is the following result that in particular implies sharpness of the classical Fréchet-bounds.

THEOREM 2.3.1 (Sharpness of Fréchet-bounds; Ru (1981a)) *Let $(E_i, \mathcal{A}_i)$ be polish spaces, $P_i \in M^1(E_i, \mathcal{A}_i)$ and $A_i \in \mathcal{A}_i$, $1 \le i \le n$; then for any $P \in M(P_1, \dots, P_n)$ holds*

$$\left(\sum_{i=1}^n P_i(A_i) - (n-1)\right)_+ \le P(A_1 \times \cdots \times A_n) \le \min\{P_i(A_i)\}$$

$$(2.3.1)$$

and the upper and lower bounds in (2.3.1) are attained.

As a consequence we get sharp bounds for the influence of dependence. As a first example, we consider the maximal risk of the components. Let $X = (X_1, \dots, X_n)$ be a random vector, with $X_i \sim P_i$ being real rv's with df's F_i. Then with $A_i = (-\infty, t]$, (2.3.1) implies sharp bounds for the maxima $M_n = \max_{i\le i\le n} X_i$.

COROLLARY 2.3.1 (Maximally dependent rv's)

$$H^-(t) = \left(\sum_{i=1}^n F_i(t) - (n-1)\right)_+ \le P\left(\max_{i\le n} X_i \le t\right)$$

$$\le \min_{1\le i\le n} F_i(t) = H^+(t) \quad (2.3.2)$$

REMARK 2.3.1

a) Corollary 2.3.1 is due for $F_1 = \cdots = F_n$ to Lai and Robbins (1976). The general case is from Lai and Robbins (1978) and by different methods from Meilijson and Nadas (1979), Tchen (1980), and Ru (1980a). Also, a random vector $\tilde{X} = (\tilde{X}_1, \dots, \tilde{X}_n)$ is constructed with $\tilde{X}_i \sim F_i$ and

$$M_n(\tilde{X}) = \max_{i\le n} \tilde{X}_i \sim H^-. \quad (2.3.3)$$

$\widetilde{X}$ yields the lower bound in (2.3.2). It is called the maximally dependent random vector in Lai and Robbins (1976). The upper bound $H^+(t)$ is attained by the comonotonic vector $X^* = (F_1^{-1}(U), \ldots, F_n^{-1}(U))$, where U is uniform on $[0, 1]$. In stochastic ordering terms (2.3.2) is equivalent to

$$M_n(X^*) \leq_{st} M_n(X) = \max_{i \leq n} X_i \leq_{st} M_n(\widetilde{X}) \quad (2.3.4)$$

Strongly positive dependent rv's have in stochastic order small maxima; i.e., they have small maximal risks of the components.

b)   The simplest way to explain the upper bound in (2.3.4) is the following argument from Lai and Robbins (1976). In fact, this is a typical argument for the duality approach to problems of this kind [see Rachev and Ru (1998)]. Note that for any real $v \in \mathbb{R}$:

$$M_n(X) = \max_{i \leq n} X_i \leq v + \sum_{i=1}^{n}(X_i - v)_+. \quad (2.3.5)$$

Equality holds in (2.3.5) iff for some "splitting point" v^* the sets $\{X_i \geq v^*\}$ are pairwise disjoint and $\bigcup_{i=1}^{n} \{X_i \geq v^*\} = \Omega$. The maximally dependent random vector $\widetilde{X}$ is constructed such that there is a splitting point v^* as above. In the case $F_1 = \cdots = F_n$, Lai and Robbins (1976) proved the extremely interesting result that the maximally dependent case is close to the independent case in the following asymptotic sense under the usual domain of attraction conditions for maxima ($\sim$ denote here asymptotic equivalence).

$$E M_n(X^{\perp}) \sim E M_n(\widetilde{X}) \sim F^{-1}\left(1 - \frac{1}{n}\right), \quad (2.3.6)$$

where $X^{\perp}$ is an *iid* sequence with df F and $a_n = F^{-1}(1 - \frac{1}{n})$ is the usual normalization for the maximum law.

There are many alternative applications of (2.3.1), e.g., to get sharp bounds for the concentration probabilities or to get sharp multivariate Fréchet-bounds [see Ru (2004)].

COROLLARY 2.3.2

a) **Maximal concentration**

$$\left(\sum_{i=1}^{n}(F_i(b_i) - F_i(a_i)) - (n-1)\right)_{+} \leq P\left(X_i \in [a_i, b_i], \ 1 \leq i \leq n\right)$$

$$\leq \min_{i \leq i \leq n} (F_i(b_i) - F_i(a_i))$$

$$(2.3.7)$$

The bounds in (2.3.7) are sharp.

b) **(Sharp) multivariate Fréchet-bounds.** *If X_i are k_i-dimensional random vectors with df's F_i, $1 \leq i \leq n$, and F is the df of $X = (X_1, \ldots, X_n)$, then for any $x_i \in \mathbb{R}^{k_i}$, $1 \leq i \leq n$:*

$$\left(\sum_{i=1}^{n} F_i(x_i) - (n-1)\right)_{+} \leq F(x_1, \ldots, x_n) \leq \min_{1 \leq i \leq n} (F_i(x_i))$$

$$(2.3.8)$$

and the multivariate Fréchet-bounds are sharp.

Of particular interest in risk theory are the distribution and risk of the combined portfolio given by the sum $S_n(X) = \sum_{i=1}^{n} X_i$.

The following basic ordering result for the ordering of sums has been stated first in Meilijson and Nadas (1979) for the convex increasing order and in Ru (1983) for the convex order.

THEOREM 2.3.2 (Maximal sums w.r.t. convex order) *Let X be a random vector with marginal df's $F_1, \ldots, F_n$; then:*

a) **Convex increasing order**

$$E\left(\sum_{i=1}^{n} X_i - t\right)_{+} \leq$$

$$\psi_{+}(t) := \inf_{v=(v_1,\ldots,v_n)} \left\{ \left(\sum v_i - t\right)_{+} + \sum_{i=1}^{n} E(X_i - v_i)_{+} \right\}$$

$$(2.3.9)$$

The bound in (2.3.9) is sharp.

b) **Convex order**

$$\sum_{i=1}^{n} X_i \prec_{cx} \sum_{i=1}^{n} F_i^{-1}(U) \tag{2.3.10}$$

$$and \ E \left(\sum_{i=1}^{n} F_i^{-1}(U) - t \right)_+ = \psi_+(t).$$

REMARK 2.3.2

1) The proof of a) was given by Meilijson and Nadas (1979) by a duality argument similar to that in (2.3.5). Also, a construction of a $r\,v$ attaining the upper bound is given there (for an even more general situation). The result of Meilijson and Nadas (1979) describes a sharp upper bound for the ordering $\prec_{icx}$ $w.r.t.$ increasing convex functions which is also called stop-loss ordering, $\prec_{sl}$ (in particular in the economics and insurance literature). That the comonotone case yields the maximum $w.r.t.$ the convex order in (2.3.10) was stated in Ru (1983) as a consequence of a more general result for supermodular functions and based on the rearrangement method which in the discrete case goes back to inequalities of Lorentz (1953). Implicitly, this result is also contained in the "Lorentz Theorem" of Tchen (1980, Theorem 5), observing that for φ convex, $\varphi(x_1 + \cdots + x_n)$ is quasimonotone (in Tchen's terminology) or supermodular in the now more common terminology. This convex ordering result for sums of random variables and also the simple duality proof have been detected and rederived several times in the literature.

2) The sharpness of the bound in (2.3.9) resp. in (2.3.10) also implies that

$$E \left(\sum_{i=1}^{n} F_i^{-1}(U) - t \right)_+$$

$$= \psi_+(t) = \inf \left\{ \sum_{i=1}^{n} E(X_i - v_i)_+; \ \sum_{i=1}^{n} v_i = t \right\}.$$

$$\tag{2.3.11}$$

If X^*, v^* attain the upper bound in (2.3.9), then

$$\left(\sum_{i=1}^{n} X_i^* - t\right)_+ = \sum_{i=1}^{n} (X_i^* - v_i^*)_+ \quad \text{a.s.} \quad (2.3.12)$$

This equality has a simple geometric meaning and is fulfilled only for the comonotonic vector $(F_i^{-1}(U))_{1 \le i \le n}$ [see Meilijson and Nadas (1979) or the recent paper by Kaas et al. (2001)].

3) Meilijson and Nadas (1979), in fact, gave sharp bounds for more general functionals. Let $I_j \subset \{1, \ldots, n\}$, $1 \le j \le k$ be subsets with $\bigcup_{j=1}^{k} I_j = \{1, \ldots, n\}$ and consider $M = \max_{1 \le i \le k} \sum_{j \in I_i} X_j$. Then for all x:

$$E(M - x)_+ \le$$

$$\widetilde{\psi_+}(x) := \inf_v \left\{ \left(\max_{1 \le i \le k} \sum_{j \in I_i} v_j - x \right)_+ \right.$$

$$\left. + \sum_{i=1}^{n} E\,(X_i - v_i)_+ \right\} \quad (2.3.13)$$

and the upper bound in (2.3.13) is pointwise sharp. Furthermore, for cyclic directed networks the bound is attained stochastically for some $P \in M(P_1, \ldots, P_n)$.

4) **Comonotonic vectors, multivariate marginals.** While for one-dimensional marginals comonotonic vectors maximize the risk for many convex functionals [like in (2.3.9), (2.3.10) for $\varphi(\sum_{i=1}^{n} x_i)$, φ convex] this is no longer the case for multivariate marginals, where they can even minimize the risk. For some illustrative examples, see Ru (2003). The reason for this is the possible negative dependence in the components of the marginals.

In the recent paper of Denuit, Dhaene, and Ribas (2001), the following interesting result related to (2.3.9), (2.3.10) was proved by a simple induction argument:

THEOREM 2.3.3 (Positive dependence increases risk) *If X is associated, then*

$$\sum_{i=1}^{n} X_i^{\perp} \leq_{sl} \sum_{i=1}^{n} X_i. \qquad (2.3.14)$$

Here $X^{\perp} = (X_1^{\perp}, \ldots, X_n^{\perp})$ has independent components and $X_i^{\perp} \sim X_i$.

Thus, positive dependence (association) leads to riskier portfolios. In Christofides and Vaggelatou (2004) and Ru (2004), a general version of this result has been given stating that positive dependence leads to higher risk for a general class of proper risk functions $f(X_1, \ldots, X_n)$.

Of particular interest in risk theory is to describe the influence of dependence on the value at risk functional of the combined portfolio $\mathrm{VaR}_\alpha(X_1 + \cdots + X_n)$ which is defined as the α-quantile of the combined portfolio $X_1 + \cdots + X_n$. For the description of the maximal influence, the following functionals are of interest: Given n df's $F_1, \ldots, F_n$ consider:

$$M_n(t) = \sup \left\{ P \left(\sum_{i=1}^{n} X_i \leq t \right); \quad X_i \sim F_i, \quad 1 \leq i \leq n \right\}$$

$$m_n(t) = \inf \left\{ P \left(\sum_{i=1}^{n} X_i < t \right); \quad X_i \sim F_i, \quad 1 \leq i \leq n \right\}.$$

$$(2.3.15)$$

Then

$$1 - m_n(t) = \sup \left\{ P \left(\sum X_i \geq t \right); \; X_i \sim F_i, \; 1 \leq i \leq n \right\}$$

$$(2.3.16)$$

and one obtains the sharp upper bound

$$\mathrm{VaR}_\alpha(X_1 + \cdots + X_n) \leq (1 - m_n)^{-1}(\alpha). \qquad (2.3.17)$$

For the case $n = 2$, the following bounds were first established in Sklar (1973) and in more general form in Moynihan, Schweizer, and Sklar (1978). The bounds and also their sharpness were independently established in Makarov (1981) and

Ru (1982). [For the history of this result, see also Schweizer (1991).]

THEOREM 2.3.4 (Maximal sum risks, $n = 2$; Makarov (1981), Ru (1982), Sklar (1973)) *Let X be a random vector with marginal df's $F_1, \ldots, F_n$; then for $n = 2$ holds:*

$$P(X_1 + X_2 \leq t) \leq M_2(t) = F_1 \wedge F_2(t)$$
$$P(X_1 + X_2 < t) \geq m_2(t) = F_1 \vee F_2(t) - 1, \qquad (2.3.18)$$

where $F_1 \wedge F_2(t) = \inf_x(F_1(x_-) + F_2(t - x))$ is the infimal convolution function and $F_1 \vee F_2(t) = \sup_x(F_1(x_-) + F_2(t-x))$ is the supremal convolution function.

(2.3.18) is derived in Ru (1982) as a consequence of the following general representation of the upper Fréchet-bounds for $\varphi = 1_A$, $P_1, P_2 \in M^1(\mathbb{R}^n)$ and $A \subset \mathbb{R}^{2n}$ closed:

$$M(A) = \sup\{P(A); \ P \in M(P_1, P_2)\}$$
$$= 1 - \sup\{P_2(O) - P_1(\pi_1(A \cap (\mathbb{R}^n \times O))); \ O \subset \mathbb{R}^n \text{ open}\},$$
$$(2.3.19)$$

where π_1 is the projection on the first component. (2.3.19) is a consequence of Strassen (1965, Theorem 11) [see Ru (1982, 1986)].

The type of bounds in (2.3.18) extends easily to $n \geq 3$ [see Frank, Nelsen, and Schweizer (1987) and Denuit, Genest, and Marceau (1999)].

PROPOSITION 2.3.1 *Let X be a random vector with marginal df's $F_1, \ldots, F_n$. Then for any $t \in \mathbb{R}^1$ holds:*

$$\left(\bigvee_{i=1}^n F_i(t) - (n-1) \right)_+ \leq P\left(\sum_{i=1}^n X_i \leq t \right)$$
$$\leq \min\left(\bigwedge_{i=1}^n F_i(t), 1 \right) \qquad (2.3.20)$$

where $\quad \bigwedge_{i=1}^{n} F_i(t) := \inf\left\{\sum_{i=1}^{n} F_i(u_i); \sum_{i=1}^{n} u_i = t\right\}$

and $\quad \bigvee_{i=1}^{n} F_i(t) := \sup\left\{\sum_{i=1}^{n} F_i(u_i); \sum_{i=1}^{n} u_i = t\right\}.$

PROOF. The proof of (2.3.20) follows by induction from the case $n = 2$ and, alternatively, by the following simple argument. For any $u_1, \ldots, u_n$ with $\sum_{i=1}^{n} u_i = t$ holds:

$$P\left(\sum_{i=1}^{n} X_i \le t\right) \le P\left(\bigcup_{i=1}^{n} \{X_i \le u_i\}\right)$$

$$\le \sum_{i=1}^{n} F_i(u_i), \tag{2.3.21}$$

which gives the upper bound. Similarly, using the Fréchet lower bound in (2.1.4), we obtain

$$P\left(\sum_{i=1}^{n} X_i \le t\right) \ge P\left(X_1 \le u_1, \ldots, X_n \le u_n\right)$$

$$\ge \left(\sum_{i=1}^{n} F_i(u_i) - (n-1)\right)_+. \tag{2.3.22}$$

$\blacksquare$

REMARK 2.3.3 The bounds in (2.3.20) are, however, in contrast to the case $n = 2$ not sharp. If $n = 3$, $F_1 = F_2 = F_3$ are the *df* of the uniform distribution on $[0, 1]$; then

$$M_3(t) = \begin{cases} \frac{2}{3}t, & 0 \le t \le \frac{3}{2} \\ 1, & t > \frac{3}{2} \end{cases}, \quad m_3(t) = \begin{cases} \frac{2}{3}t - 1, & 0 \le t \le 3 \\ 1, & t \ge 3 \end{cases}$$

$$\tag{2.3.23}$$

[see Ru (1982)]. The bounds in (2.3.20) and crude (2.3.21) are in this case.

$$\min\left(1, \bigwedge_{i=1}^{3} F_i(t)\right) = \min(1, t) \text{ and } \left(\bigvee_{i=1}^{3} F_i(t) - 2\right)_+$$

$$= (t - 2)_+. \tag{2.3.24}$$

For some examples sharp bounds for $n \geq 3$ have been given in Ru (1982) and Rachev and Ru (1998).

The simple method of bounding the risk probability in (2.3.20) has been given and extended in Frank, Nelsen, and Schweizer (1987) to general monotonically nondecreasing functions $\psi(x_1, \ldots, x_n)$. The resulting bounds are of interest and markable relevance if further information on the underlying df's can be used. The following is essentially a reformulation of corresponding results in Moynihan, Schweizer, and Sklar (1978); Frank, Nelsen, and Schweizer (1987); Denuit, Genest, and Marceau (1999); and Embrechts, Höing, and Juri (2003). For a df H, let $\bar{H}$ denote the corresponding multivariate survival function $\bar{H}(x) = P_H([x, \infty))$. For $t \in \mathbb{R}$, let $A_\psi^+(t) := \{u = (u_1, \ldots, u_n) : u$ a maximal point in $\mathbb{R}^n$ with $\psi(u) \leq t\}$.

THEOREM 2.3.5 (Bounds for monotonic functionals) *Let* $X = (X_1, \ldots, X_n)$ *be a random vector with* df $F \in \mathcal{F}(F_1, \ldots, F_n)$ *and let* $\psi(x)$ *be monotonically nondecreasing and lower semi-continuous. Then*

a) **General bounds**

$$\left(\sup_{u \in A_\psi^+(t)} \sum_{i=1}^n F_i(u_i) - (n-1) \right)_+ \leq P\left(\psi(X) \leq t\right)$$

$$\leq \inf_{u \in A_\psi^+(t)} \sum_{i=1}^n F_i(u_i). \quad (2.3.25)$$

b) **Improved bounds.** *If* G, H *are* df's, *then*
   1) $F \geq G$ *implies*

$$P\left(\psi(X) \leq t\right) \geq \sup_{u \in A_\psi^+(t)} G(u). \quad (2.3.26)$$

   2) *If* $\bar{F} \geq \bar{H}$, *then*

$$P\left(\psi(X) < t\right) \leq 1 - \sup_{u \in A_\psi^-(t)} \bar{H}(u), \quad (2.3.27)$$

   where $A_\psi^-(t) := \{u \in \mathbb{R}^n : \psi(u) \geq t\}$.

PROOF.

a) For any $u \in A_\psi^+(t)$ holds, using maximality of u,

$$P\left(\psi(X) \le t\right) \le P\left(\bigcup_{i=1}^{n}\{X_i \le u_i\}\right) \le \sum_{i=1}^{n} F_i(u_i).$$

This implies the upper bound in (a). Further,

$$P\left(\psi(X) \le t\right) \ge P\left(X_1 \le u_1, \ldots, X_n \le u_n\right)$$

$$\ge \left(\sum_{i=1}^{n} F_i(u_i) - (n-1)\right)_+ \qquad (2.3.28)$$

by the lower Fréchet-bound.

b) If $F \ge G$, then in (2.3.28) we get

$$P\left(\psi(X) \le t\right) \ge \left(\sup_{u \in A_\psi^+(t)} G(u) - (n-1)\right)_+.$$

If $\overline{F} \ge \overline{H}$ then for $u \in A_\psi^-(t)$

$$P\left(\psi(X) < t\right) = 1 - P\left(\psi(X) \ge t\right)$$

$$\le 1 - P\left(X_1 \ge u_1, \ldots, X_n \ge u_n\right)$$

$$= 1 - \overline{F}(u) \le 1 - \overline{H}(u),$$

which implies (2). ∎

REMARK 2.3.4

a) For the case $n = 2$ one gets sharp upper and lower bounds for $P(\psi(X) \ge t)$ by applying (2.3.20) to the set $A = \{x = (x_1, x_2) : \psi(x_1, x_2) \ge t\}$ for any function ψ, in particular, for monotonically nondecreasing functions.

b) Theorem 3.2 in Embrechts, Höing, and Juri (2003) states sharpness of the bounds in (2.3.26) and (2.3.27). In comparison to Embrechts, Höing, and Juri (2003), we omit some continuity assumption on ψ and omit the language of copulas which is not necessary here. The corresponding bound for the value at risk functionals are in consequence of the monotonicity of ψ easy to achieve [see Embrechts et al. (2003, Theorem 4.1)]. There are still many open problems in this area, in particular how to obtain applicable and good bounds under additional information on the model.

c) An extension of the bounds in (2.3.14), (2.3.25) to increasing functions $\psi(X,Y)$ of k-dimensional vectors has been given in Li, Scarsini, and Shaked (1996). For $n=2$ one gets sharpness by Strassen's theorem as in (2.3.17). Also, the partial integration argument from Ru (1980a) can be applied to obtain bounds for $Eg(X+Y)$ for increasing differentiable functions g [see Li, Scarsini, and Shaked (1996, Theorem 4.2)] and more general to any Δ-monotone functions $f(X_1,\ldots,X_n)$ for k_i-dimensional random vectors X_i [see Ru (2004)]. Several further bounds and techniques for obtaining bounds are discussed in Ru (1991a). One rich source of such bounds comprises Bonferroni inequalities which in many cases can be proved by a general reduction principle to be sharp. Let, e.g., $A_1,\ldots,A_n \in \mathcal{A}$ where $(E,\mathcal{A})$ is any measure space and $P_i \in M^1(E,\mathcal{A})$. Let $X=(X_1,\ldots,X_n)$ be a random vector with $X_j \sim P_j$, $1 \le j \le n$ and define the set that at least k of the events $\{X_j \in A_j\}$ hold true,

$$L_k := \bigcup_{J \subset \{1,\ldots,n\},\ |J|=k} \{X_j \in A_j,\ j \in J\}. \quad (2.3.29)$$

Then

$$P(L_k) \le b_k := \min_{0 \le r \le k-1} \left(1, \frac{1}{k-r}\sum_{i=1}^{n-r} p_{(i)}\right)$$

$$(2.3.30)$$

$$P(L_k) \ge a_k := \max\left(0, \frac{\sum_{i=r+1}^{n} p_{(i)} - (k-1)}{n-r-(k-1)}\right)$$

where $p_i = P(X_i \in A_i)$ and $p_{(1)} \le \cdots \le p_{(n)}$ [see Ru (1991a)]. The bounds in 2.3.30 are sharp. They are consequences of a Bonferroni type result in Rüger (1979) and a general reduction principle [see Ru (1991a)]. In particular, for real random variables and

$A_i = [t, \infty)$, one gets sharp upper and lower bounds for the tail of the kth-order statistic

$$P\left(X_{(k)} \geq t\right) \begin{cases} \leq b_k, \text{ with } p_i = P(X_i \geq t), \\ \geq a_k. \end{cases} \quad (2.3.31)$$

Also extensions to higher-order Bonferroni bounds are given in Ru (1991) and to improved bounds in the case that one can use some of the higher-order joint marginal distributions.

2.4 Δ-MONOTONE, SUPERMODULAR, AND DIRECTIONALLY CONVEX FUNCTION CLASSES

In various applications of comparing risks, it has turned out that Δ-monotone, supermodular, and directionally convex functions and variants of them play an eminent role [see Müller and Stoyan (2002)]. For the definition we introduce for $f: \mathbb{R}^n \to \mathbb{R}$, $\epsilon > 0$, the difference operator $\Delta_i^\epsilon f$ by

$$\Delta_i^\epsilon f(x) = f(x + \epsilon e_i) - f(x), \quad 1 \leq i \leq n \quad (2.4.1)$$

where e_i is the ith unit vector.

DEFINITION 2.4.1 Let $f : \mathbb{R}^n \to \mathbb{R}$.

1) f is Δ-monotone if for every subset $I = \{i_1, \ldots, i_k\} \subset \{1, \ldots, n\}$ and $\epsilon_1, \ldots, \epsilon_k > 0$ holds

$$\Delta_{i_1}^{\epsilon_1} \ldots \Delta_{i_k}^{\epsilon_k} f(x) \geq 0 \quad \text{for all } x. \quad (2.4.2)$$

2) f is supermodular if for all $1 \leq i < j \leq n$, $\epsilon, \delta > 0$ and all x

$$\Delta_i^\epsilon \Delta_j^\delta f(x) \geq 0. \quad (2.4.3)$$

3) f is directionally convex if (2.4.3) holds for all $i \leq j$.

Denote by $\mathcal{F}^\Delta$, $\mathcal{F}^{sm}$, $\mathcal{F}^{dcx}$ the set of all Δ-monotone resp. supermodular resp. directionally convex functions; then $\mathcal{F}^\Delta \subset \mathcal{F}^{sm}$ and $\mathcal{F}^{dcx} \subset \mathcal{F}^{sm}$.

REMARK 2.4.1 The classes of supermodular and directionally convex functions were investigated in early papers of

Lorentz (1953) and Fan and Lorentz (1954) in the context of functional inequalities [see also the extensive chapter in Marshall and Olkin (1979)]. In Cambanis, Simons, and Stout (1976) and Tchen (1980), supermodular functions are called quasimonotone. Δ-monotone functions were introduced in Ru (1980). Twice-differentiable functions f are supermodular (directionally convex) if

$$\frac{\partial^2}{\partial x_i \partial x_j} f(x) \geq 0 \quad \text{for all } x \text{ and } i < j \text{ (resp. for } i \leq j\text{)}.$$

(2.4.4)

Differentiable functions f are Δ-monotone if for all $i_1 < i_2 < \cdots < i_k, 1 \leq k \leq n$

$$\frac{\partial^k}{\partial x_{i_1} \ldots \partial x_{i_k}} f(x) \geq 0. \tag{2.4.5}$$

DEFINITION 2.4.2 For $P, Q \in M^1(\mathbb{R}^n)$ define

a) $P \leq_{uo} Q$ "upper orthant ordering" if

$$P([x, \infty]) \leq Q([x, \infty]), \quad \forall x \in \mathbb{R}^n. \tag{2.4.6}$$

b) $\leq_{sm}, \leq_{dcx}$ denote the supermodular ordering resp. directionally convex ordering generated by $\mathcal{F}^{sm}$ resp. $\mathcal{F}^{dcx}$.

There are corresponding positive/negative dependence notions.

DEFINITION 2.4.3

1) P is positive (negative) upper orthant dependent— $P \in$ PUOD (resp. $P \in$ NUOD)—if

$$\bigotimes_{i=1}^n P_i \leq_{uo} P \left(\text{resp. } P \leq_{uo} \bigotimes_{i=1}^n P_i \right). \tag{2.4.7}$$

2) P is weakly associated if $E \prod_{i=1}^n f_i(X_i) \geq \prod_{i=1}^n E f_i(X_i)$ for all nondecreasing $f_i \geq 0$.

REMARK 2.4.2 A similar notion also exists for lower orthants and is denoted by $\leq_{lo}$ resp. PLOD and NLOD. This notion was introduced in Lehmann (1966). The following equivalence holds:

$P \in$ PUOD if and only if P is weakly associated   (2.4.8)

[for $n = 2$ due to Lehmann (1966), for $n \geq 2$ to Ru (1981c)]. In fact, more generally it was shown in Bergmann (1978) that: $P \leq_{uo} Q$ if and only if for $X \sim P, Y \sim Q$

$$E \prod_{i=1}^{n} f_i(X_i) \leq E \prod_{i=1}^{n} f_i(Y_i), \tag{2.4.9}$$

for f_i nondecreasing, $f_i \geq 0$.

The maximal generator of the upper orthant order is the set of Δ-monotone functions.

THEOREM 2.4.1 (Δ-monotone functions; Ru (1980a)) *If $P, Q \in M^1(\mathbb{R}^n)$, then: $P \leq_{uo} Q$ if and only if*

$$\int f \, dP \leq \int f \, dQ \quad \text{for all } f \in \mathcal{F}^{\Delta} \tag{2.4.10}$$

which are integrable w.r.t. P and Q, i.e., $\leq_{uo}$ is equivalent to $\leq_{\mathcal{F}^{\Delta}}$.

A similar result, of course, also holds for the lower orthant order and can be combined to characterize the "concordance ordering"; $P \leq_{con} Q$ if $P \leq_{uo} Q$ and $P \leq_{lo} Q$. A random vector X is called WA (weakly associated) if $\mathrm{Cov}(f(X_J), g(X_I)) \geq 0$ for any disjoint subsets J, I of $\{1, \ldots, n\}$ and monotonically nondecreasing functions f, g of these components. Concerning the supermodular ordering the analogous result is the following.

THEOREM 2.4.2 (Supermodular functions) *Let $P, Q \in M^1(\mathbb{R}^n)$,*

a) $n = 2$; *Cambanis, Simons, and Stout (1976). For $P, Q \in M(P_1, P_2)$ holds:*

$$P \leq_{uo} Q \Longleftrightarrow P \leq_{sm} Q. \tag{2.4.11}$$

b) $n \geq 2$, "The Lorentz Theorem"; *Tchen (1980), Ru (1983). For $P \in M(P_1, \ldots, P_n)$ holds*

$$P \leq_{sm} P_+, \tag{2.4.12}$$

where P_+ is the measure corresponding to the upper Fréchet-bound (the comonotonic measure).

c) *Christofides and Vaggelatou (2003).*
   If X is a weakly associated random vector, then X has
   positive supermodular dependence , i.e.,

$$\bigotimes_{i=1}^{n} P^{X_i} \leq_{sm} P^X. \tag{2.4.13}$$

REMARK 2.4.3

a) The interesting result in (2.4.13) can be stated in the
   form: Positive dependence implies increasing of the
   risk. This is of essential interest in risk theory.
b) (2.4.12) was proved in Tchen (1980) by discrete approximation and reduction to the Lorentz (1953)
   inequalities. In Ru (1979, 1983), the problem of generalized Fréchet-bounds was identified with a rearrangement problem for functions and then reduced to
   the Lorentz inequality.
c) For P and Q with identical $(n-1)$-dimensions marginal
   distributions, one obtains

$$P \leq_{uo} Q \Rightarrow P \leq_{sm} Q \tag{2.4.14}$$

   [Tchen (1980), Ru (1980a, Theorem 3b)].
d) There are some useful composition rules which al-
   low the use of $\leq_{sm}$ for several models of interest [see
   Müller and Stoyan (2002) for results and references].
e) From Tchen's proof of (b), it is clear that $P_- \leq_{sm} P$ if
   the lower Fréchet-bound P_- is a df [see also Müller
   and Stoyan (2002, p. 120)].
f) Since for any φ convex the function $\psi(x) = \varphi(x_1 + \cdots + x_n)$ is supermodular, one obtains as a consequence of
   (2.4.12) the statement of (2.3.9) that

$$\sum_{i=1}^{n} X_i \leq_{cx} \sum_{i=1}^{n} F_i^{-1}(U) \tag{2.4.15}$$

   i.e., the sums are maximal in convex order for the
   comonotone case; in other words, the comonotone pos-
   itively dependent portfolio is the riskiest possible.
g) Comparison of $P \in M(P_1, \ldots, P_n)$ and $Q \in M(Q_1, \ldots, Q_n)$ w.r.t. the supermodular ordering $\leq_{sm}$ is

only possible if the marginals are identical:

$$P \leq_{sm} Q \text{ implies } P_i = Q_i, \quad 1 \leq i \leq n. \quad (2.4.16)$$

Thus, in comparison to $\leq_{\mathcal{F}^\Delta} = \leq_{uo}$, the $\leq_{sm}$ ordering is restricted to one marginal class while $\leq_{uo}$ allows comparisons between P and Q if the marginals increase stochastically:

$$P \leq_{uo} Q \text{ implies } P_i \leq_{st} Q_i, \quad 1 \leq i \leq n. \quad (2.4.17)$$

The comparison by $\mathcal{F}^\Delta$ is, however, for a smaller class of functions $\mathcal{F}^\Delta \subset \mathcal{F}^{sm}$. On the other hand, criteria for $\leq_{sm}$ are not as simple as those for $\leq_{uo}$. The directionally convex order $\leq_{dcx}$ is a "typical" risk order. It allows comparisons in cases where the marginals increase convexly:

$$P \leq_{dcx} Q \text{ implies } P_i \leq_{cx} Q_i, \quad 1 \leq i \leq n. \quad (2.4.18)$$

From the copula representation (2.2.7) of distributions with given marginals, the following is immediate: If $P \in M$ $(P_1, \ldots, P_n)$, $Q \in M(Q_1, \ldots, Q_n)$ have the same copula C and $P_i \leq_{st} Q_i$, $1 \leq i \leq n$, then

$$P \leq_{st} Q. \quad (2.4.19)$$

($\leq_{st}$ is the multivariate stochastic order w.r.t. increasing function; see Ru (1981b, Proposition 7).)

The situation is more complicated if the marginals increase in convex order. Here the analog of (2.4.19) is wrong; see Müller and Scarsini (2001). The reason is that negative dependence can destroy this conclusion, as the following simple example of that paper shows.

EXAMPLE 2.4.1 Consider $n = 2$ and rv's $X = (W, -W)$, $Y = (W, -EW)$ for some integrable random variable W. Then $Y_i \leq_{cx} X_i$, $i = 1, 2$, but $X_1 + X_2 = W - W = 0$ while $Y_1 + Y_2 = W - EW$, i.e., $X_1 + X_2 \leq_{cx} Y_1 + Y_2$.

However, in the positive direction Müller and Scarsini (2001) proved the interesting result that under a strong positive dependence assumption the analog of (2.4.19) is true. Let F_+ denote as usual the upper Fréchet-bound of $\mathcal{F}(F_1, \ldots, F_n)$.

THEOREM 2.4.3 (Directionally convex ordering) *Let F_i, G_i be one-dimensional df's, $1 \leq i \leq n$.*

    a) *"The Ky-Fan-Lorentz Theorem"; Ru (1983). If $F_i \leq_{cx} G_i$, $1 \leq i \leq n$, then*

$$F_+ \leq_{dcx} G_+. \tag{2.4.20}$$

    b) *Müller and Scarsini (2001). If $F \in \mathcal{F}(F_1, \ldots, F_n)$ and $G \in \mathcal{F}(G_1, \ldots, G_n)$ have the same conditionally increasing (CI) copula C and if $F_i \leq_{cx} G_i$, $1 \leq i \leq n$, then*

$$F \leq_{dcx} G. \tag{2.4.21}$$

REMARK 2.4.4 Müller and Scarsini (2001) give a proof of (a) using mean preserving spread (see Theorem 3.12.13 of their paper), whereas the proof in Ru (1983) is based on the Ky Fan and Lorentz Theorem. The second main ingredient of the proof of (b) is the a.s. standard construction of random vectors in (2.2.4): $X = \tau_F^*(V) = (h_1(V_1), \ldots, h_n(V_1, \ldots, V_n))$, where the functions h_i are monotonically nondecreasing for CI distribution functions. (2.4.21) is not valid anymore under the weaker dependence assumption of association or of conditional increasing in sequence CIS [see Müller and Scasini (2001)].

The following weakening of the WA-notion was introduced in Ru (2003):

X is smaller than Y in the *weakly conditional in sequence order—$X \leq_{WCS} Y$*—if, for all t, $1 \leq i \leq n-1$ and f monotonically nondecreasing:

$$\mathrm{Cov}(1(X_i > t), f(X_{(i+1)}))$$
$$\leq \mathrm{Cov}(1(Y_i > t), f(Y_{(i+1)})) \tag{2.4.22}$$

where $X_{(i+1)} = (X_{i+1}, \ldots, X_n)$. X is called *weakly associated in sequence (WAS)* if $X^* \leq_{WCS} X$, where X^* is the corresponding version of X with independent components; equivalently, for all t,

$$P^{X_{(i+1)}|X_i > t} \geq_{st} P^{X_{(i+1)}}. \tag{2.4.23}$$

The following result extends and unifies Theorems 2.4.2 and 2.4.3.

THEOREM 2.4.4 (WCS-Theorem; Ru (2004)) *Let X, Y be random vectors with marginals P_i, Q_i.*

a) *If $P_i = Q_i, 1 \leq i \leq n$ and $X \leq_{WCS} Y$, then $X \leq_{sm} Y$.*
b) *If $P_i \leq_{cx} Q_i, 1 \leq i \leq n$ and $X \leq_{WCS} Y$, then $X \leq_{dcx} Y$.*

The ordering $\leq_{WCS}$ combines an increase in positive dependence with a convex increase of the marginals. Some examples for this ordering are given in Ru (2004). In particular, one obtains as a corollary:

COROLLARY 2.4.1 *If $F \in \mathcal{F}(F_1, \ldots, F_n)$ and $F_i \leq_{cx} G_i, 1 \leq i \leq n$, then*

a) $$F \leq_{dcx} G_+. \tag{2.4.24}$$

b) *If $X \sim F$, then*

$$\sum_{i=1}^{n} X_i \leq_{cx} \sum_{i=1}^{n} G_i^{-1}(U), \tag{2.4.25}$$

where U is uniformly distributed on $(0, 1)$.

Finally, we state some new criteria for the $\leq_{sm}$ and $\leq_{dcx}$ ordering in functional dependence models which are related to Bäuerle (1997, Theorem 3.1) and Bäuerle and Müller (1998). For the proofs see Ru (2004). Let (U_i) be independent rv's and (V_i), V any random variables independent of (U_i). Further, let

$$X_i = g_i(U_i, V_i), \; Y_i = g_i(U_i, V),$$
$$\tag{2.4.26}$$
$$Z_i = \tilde{g}_i(U_i, V_i), \; W_i = \tilde{g}_i(U_i, V)$$

where $V_i \sim V$ and $g_i(u, \cdot), \tilde{g}_i(u, \cdot)$ are monotonically nondecreasing. Let $X = (X_1, \ldots, X_n), Y = (Y_1, \ldots, Y_n), Z$, and W denote the corresponding vectors and let $\leq_{ccx}$ denote the componentwise convex order. X, Y, Z, and W describe functional models where the dependence is obtained in functional form from some inner and outer factors U_i resp. V_i. These types of models are of particular relevance in various applications in insurance and in economics.

THEOREM 2.4.5 *For the X, Y, Z, and W specified as in (2.4.26) holds:*

  a) *Bäuerle (1997). $X \leq_{sm} Y$, $Z \leq_{sm} W$.*
  b) *If for all v, $\widetilde{g}_i(U_i, v) \leq_{cx} g_i(U_i, v)$, then $Z \leq_{ccx} X$, $W \leq_{ccx} Y$, and $Z \leq_{dcx} Y$.*
  c) *If $g_i(U_i, v) \leq_{cx} \widetilde{g}_i(U_i, v)$, then $X \leq_{ccx} Z$, $Y \leq_{ccx} W$ and $X \leq_{dcx} W$.*

For the proofs of (b) and (c) see Ru (2004).

REMARK 2.4.5 The random vectors Z, Y and X, W which are compared w.r.t. $\leq_{dcx}$ in Theorem 2.4.5 do not have the same dependence structure (copula), which was a basic assumption for the proof of the $\leq_{dcx}$ ordering result in Theorem 2.4.3 (b). Also, the X and W vectors are not necessarily positive dependent. Since we do not postulate any independence for the (V_i), we can describe any multivariate df F by a random vector of the form as for X. Thus, this comparison result applies to many models. Similarly, as in Bäuerle (1997), one could add in Theorem 2.4.5 a further random influence component W and consider models of the form $X_i = g_i(U_i, V_i, W)$, $Y_i = g_i(U_i, V_i, W)$, etc.

REFERENCES

Bäuerle, N. (1997). Inequalities for stochastic models via supermodular orderings. *Communication in Statistics—Stochastic Models*, **13**, 181–201.

Bäuerle, N. and Müller, A. (1998). Modeling and comparing dependencies in multivariate risk portfolios. *ASTIN Bulletin*, **28**, 59–76.

Beneš, V. and Štěpán, J. (Eds.) (1997). *Distributions with Given Marginals and Moment Problems*, Kluwer Academic Publishers.

Bergmann, R. (1978). Some classes of semi-ordering relations for random vectors and their use for comparing covariances. *Mathematische Nachrichten*, **82**, 103–114.

Bergmann, R. (1991). Stochastic orders and their application to a unified approach to various concepts of dependence and association. In *Stochastic Orders and Decision under Risk* (Eds., K. Mosler and M. Scarsini), pp. 48–73, Volume 19 of *IMS Lecture Notes— Monograph Series*, Institute of Mathematical Statistics, Hayward, California.

Cambanis, S., Simons, G. and Stout, W. (1976). Inequalities for $Ek(X, Y)$ when the marginals are fixed. *Zeitschrift für Wahrscheinlichkeitstheorie und verwandte Gebiete*, **36**, 285–294.

Chong, K.-M. (1974). Some extensions of a theorem of Hardy, Littlewood and Polya and their applications. *Canadian Journal of Mathematics*, **26**, 1321–1340.

Christofides, C. and Vaggelatou, E. (2004). A connection between supermodular ordering and positive/negative association. *Journal of Multivariate Analysis*, **88**, 138–151.

Cohen, A. and Sackrowitz, B. H. (1995). On stochastic ordering of random vectors. *Journal of Applied Probability*, **32**, 960–965.

Cuadras, C. M., Fortiana, J. and Rodríguez-Lallena, J. A. (Eds.) (2002). *Distributions with Given Marginals and Statistical Modeling*, Kluwer Academic Publishers, Dordrecht, The Netherlands.

Dall'Aglio, G. (1972). Fréchet classes and compatibility of distribution functions. In *Symposia Mathematica*, Volume 9, pp. 131–150, American Mathematical Society, Providence, Rhode Island.

Dall'Aglio, G., Kotz, S. and Salinetti, G. (Eds.) (1991). *Advances in Probability Distributions with Given Marginals*, Kluwer Academic Publishers, Dordrecht, The Netherlands.

Day, P. W. (1972). Rearrangement inequalities. *Canadian Journal of Mathematics*, **24**, 930–943.

Denuit, M. and Müller, A. (2002). Smooth generators of integral stochastic orders. *Annals of Applied Probability*, **12**, 1174–1184.

Denuit, M., Dhaene, J. and Ribas, C. (2001). Does positive dependence between individual risks increase stop-loss premiums? *Insurance Mathematics and Economics*, **28**, 305–308.

Denuit, M., Genest, C. and Marceau, E. (1999). Stochastic bounds on sums of dependent risks. *Insurance Mathematics and Economics*, **25**, 85–104.

Dhaene, J., Wang, S., Young, V. and Goovaerts, M. J. (2000). Comonotonicity and maximal stop-loss premiums. *Bulletin of the Swiss Association of Actuaries*, **2**, 99–113.

Embrechts, P., Höing, A. and Juri, A. (2003). Using copulae to bound the value-at-risk for functions of dependent risks. *Finance & Stochastics*, **7**, 145–167.

Fan, K. and Lorentz, G. G. (1954). An integral inequality. *American Mathematical Monthly*, **61**, 626–631.

Frank, M. J., Nelsen, R. B. and Schweizer, B. (1987). Best-possible bounds for the distribution for a sum—a problem of Kolmogorov. *Probability Theory and Related Fields*, **74**, 199–211.

Franken, P. and Kirstein, B. M. (1977). Zur Vergleichbarkeit zufälliger Prozesse. *Mathematische Nachrichten*, **78**, 197–205.

Franken, P. and Stoyan, D. (1975). Einige Bemerkungen über monotone und vergleichbare markowsche Prozesse. *Mathematische Nachrichten*, **66**, 201–209.

Fréchet, M. (1951). Sur le tableaux de corrélation dont les marges sont données. *Annales de l'Université de Lyon, Section A*, **14**, 53–77.

Hoeffding, W. (1940). Masstabinvariante Korrelationstheorie. *Schriften des Mathematischen Instituts und des Instituts für Angewandte Mathematik der Universität Berlin*, **5**, 179–233.

Joag-Dev, K. and Proschan, F. (1983). Negative association of random variables, with applications. *Annals of Statistics*, **11**, 286–295.

Joe, H. (1997). *Multivariate Models and Dependence Concepts*, Volume 73 of *Monographs on Statistics and Applied Probability*, Chapman & Hall, London.

Kaas, R., Dhaene, J., Vyncke, D., Goovaerts, M. J. and Denuit, M. (2001). A simple geometric proof that comonotonic risks have the convex largest sum. In *Proceedings of the Fifth International Congress on Insurance*, Mathematics and Economics, State College.

Kalmykov, G. I. (1962). On the partial ordering of one-dimensional Markov processes on the real line. *Theory of Probability and Its Applications*, **7**, 456–459.

Kamae, T. and Krengel, U. (1978). Stochastic partial ordering. *Annals of Probability*, **6**, 1044–1049.

Kamae, T., Krengel, U. and O'Brien, G. L. (1977). Stochastic inequalities on partially ordered spaces. *Annals of Probability*, **5**, 899–912.

Kellerer, H. G. (1984). Duality theorems for marginal problems. *Zeitschrift für Wahrscheinlichkeitstheorie und verwandte Gebiete*, **67**, 399–432.

Lai, T. L. and Robbins, M. (1976). Maximally dependent random variables. *Proceedings of the National Academy of Sciences*, **73**, 286–288.

Lai, T. L. and Robbins, H. (1978). A class of dependent random variables and their maxima. *Zeitschrift für Wahrscheinlichkeitstheorie und verwandte Gebiete*, **42**, 89–112.

Lehmann, E. L. (1966). Some concepts of dependence. *Annals of Mathematical Statistics*, 1137–1153.

Li, H., Scarsini, M. and Shaked, M. (1996). Bounds for the distribution of a multivariate sum. In *Proceedings of Distributions with Fixed Marginals and Related Topics*, pp. 198–212, Volume 28 of *IMS Lecture Notes Monograph Series*, Institute of Mathematical Statistics, Hayward, California.

Lorentz, G. G. (1953). An inequality for rearrangements. *American Mathematical Monthly*, **60**, 176–179.

Makarov, G. D. (1981). Estimates for the distribution function of a sum of two random variables when the marginal distributions are fixed. *Theory of Probability and Its Applications*, **26**, 803–806.

Marshall, A. W. (1991). Multivariate stochastic orderings and generating cones of functions. In *Stochastic Orders and Decision under Risk* (Eds., K. Mosler and M. Scarsini), pp. 231–247, Volume 19 of *IMS Lecture Notes—Monograph Series*, Institute of Mathematical Statistics, Hayward, California.

Marshall, A. W. and Olkin, I. (1979). *Inequalities: Theory of Majorization and Its Applications*, Academic Press, New York.

Meilijson, I. and Nadas, A. (1979). Convex majorization with an application to the length of critical paths. *Journal of Applied Probability*, **16**, 671–677.

Meyer, P. A. (1966). *Probability and Potentials*, Blaisdell Publishing Company.

Mosler, K. and Scarsini, M. (Eds.) (1991a). *Stochastic Orders and Decision under Risk*, Volume 19 of *IMS Lecture Notes—Monograph Series*, Institute of Mathematical Statistics, Hayward, California.

Mosler, K. and Scarsini, M. (1991b). Some theory of stochastic dominance. In *Stochastic Orders and Decision under Risk* (Eds., K. Mosler and M. Scarsini), pp. 261–284, Volume 19 of *IMS Lecture Notes—Monograph Series*, Institute of Mathematical Statistics, Hayward, California.

Mosler, K. C. (1982). *Entscheidungsregeln bei Risiko: Multivariate Stochastische Dominanz*, Volume 204 of *Lecture Notes in Economics and Math. Systems*, Springer-Verlag, New York.

Moynihan, R., Schweizer, B. and Sklar, A. (1978). Inequalities among binary operations on probability distribution functions. In *General Inequalities 1* (Ed. E. F. Beckenbach), pp. 133–149, Birkhäuser Verlag, Basel.

Müller, A. (1997). Stochastic orders generated by integrals: A unified study. *Advances in Applied Probability*, **29**, 414–428.

Müller, A. and Scarsini, M. (2000). Some remarks on the supermodular order. *Journal of Multivariate Analysis*, **73(1)**, 107–119.

Müller, A. and Scarsini, M.(2001). Stochastic comparison of random vectors with a common copula. *Mathematics of Operations Research*, **26**, 723–740.

Müller, A. and Stoyan, D. (2002). *Comparison Methods for Stochastic Models and Risks*, John Wiley & Sons, Chichester, U.K.

Nelsen, R. B. (1999). *An Introduction to Copulas*, Volume 139 of *Lecture Notes in Statistics*, Springer-Verlag, New York.

O'Brien, G. L. (1975). The comparison method for stochastic processes. *Annals of Probability*, **3**, 80–88.

Rachev, S. T. and Rüschendorf, L. (1998). *Mass Transportation Problems*, Vol. I/II, Springer-Verlag, New York.

Ramachandran, D. and Rüschendorf, L. (1995). A general duality theorem for marginal problems. *Probability Theory and Related Fields*, **101**, 311–319.

Reuter, H. and Riedrich, T. (1981). On maximal sets of functions compatible with a partial ordering for distribution functions. *Mathematisch Operationsforschung und Statistik, Series Optimization*, **12**, 597–605.

Rosenblatt, M. (1952). Remarks on a multivariate transformation. *Annals of Mathematical Statistics*, **23**, 470–472.

Rüger, B. (1979). Scharfe untere und obere Schranken für die Wahrscheinlichkeit der Realisation von k unter n Ereignissen. *Metrika*, **26**, 71–77.

Rüschendorf, L. (1979). *Vergleich von Zufallsvariablen bzgl. integral-induzierter Halbordnungen*. Habilitationsschrift.

Rüschendorf, L. (1980a). Inequalities for the expectation of Δ-monotone functions. *Zeitschrift für Wahrscheinlichkeitstheorie und verwandte Gebiete*, **54**, 341–349.

Rüschendorf, L. (1980b). Ordering of distributions and rearrangement of functions. *Annals of Probability*, **9**, 276–283.

Rüschendorf, L. (1981a). Sharpness of Fréchet-bounds. *Zeitschrift für Wahrscheinlichkeitstheorie und verwandte Gebiete*, **57**, 293–302.

Rüschendorf, L. (1981b). Stochastically ordered distributions and monotonicity of the OC of an SPRT. *Mathematisch Operationsforschung und Statistik, Series Statistics*, **12**, 327–338.

Rüschendorf, L. (1981c). Weak association of random variables. *Journal of Multivariate Analysis*, **11**, 448–451.

Rüschendorf, L. (1982). Random variables with maximum sums. *Advances in Applied Probability*, **14**, 623–632.

Rüschendorf, L. (1983). Solution of statistical optimization problem by rearrangement methods. *Metrika*, **30**, 55–61.

Rüschendorf, L. (1986). Monotonicity and unbiasedness of tests via a.s. constructions. *Statistics*, **17**, 221–230.

Rüschendorf, L. (1991a). Bounds for distributions with multivariate marginals. In *Stochastic Orders and Decision under Risk* (Eds., K. Mosler and M. Scarsini), pp. 285–310, Volume 19 of *IMS Lecture Notes—Monograph Series*, Institute of Mathematical Statistics, Hayward, California.

Rüschendorf, L. (1991b). Fréchet bounds and their applications. In *Advances in Probability Distributions with Given Marginals: Beyond the Copulas* (Eds., G. Dall'Aglio, S. Kotz, and G. Salinetti), Kluwer Academic Publishers, Dordrecht, The Netherlands.

Rüschendorf, L. (1996). Developments on Fréchet bounds. In *Proceedings of Distributions with Fixed Marginals and Related Topics*, pp. 273–296, Volume 28 of *IMS Lecture Notes Monograph Series*, Institute of Mathematical Statistics, Hayward, California.

Rüschendorf, L. (2004). Comparison of multivariate risks, Fréchet-bounds, and positive dependence. *Journal of Applied Probability*, **41**, 391–406.

Rüschendorf, L., Schweizer, B. and Taylor M. D. (Eds.) (1996). *Distributions with Fixed Marginals and Related Topics*, Volume 28 of *IMS Lecture Notes Monograph Series*, Institute of Mathematical Statistics, Hayward, California.

Rüschendorf, L. and de Valk, V. (1993). On regression representations of stochastic processes. *Stochastic Processes and Their Applications*, **46**, 183–198.

Schweizer, B. (1991). Thirty years of copulas. In *Advances in Probability Distributions with Given Marginals: Beyond the Copulas* (Eds., G. Dall'Aglio, S. Kotz, and G. Salinetti), Kluwer Academic Publishers, Dordrecht, The Netherlands.

Shaked, M. and Shantikumar, J. G. (1994). *Stochastic Orders and Their Applications*. Academic Press, Boston, Massachusetts.

Shaked, M. and Shantikumar, J. G. (1997). Supermodular stochastic orders and positive dependence of random vectors. *Journal of Multivariate Analysis*, **61**, 86–101.

Shaked, M. and Tong, Y. L. (1985). Some partial orderings of exchangeable random variables by positive dependence. *Journal of Multivariate Analysis*, **17**, 333–349.

Sklar, A. (1959). *Fonctions de répartition á n dimensions et leurs marges*. pp. 229–231, Institute of Statistics, University of Paris 8, France.

Sklar, A. (1973). Random variables, joint distribution functions and copulas. *Krybernetika*, **9**, 449–460.

Stoyan, D. (1972). Halbordnungsrelationen für Verteilungsgesetze. *Mathematische Nachrichten*, **52**, 315–331.

Stoyan, D. (1977). *Qualitative Eigenschaften und Abschätzungen stochastischer Modelle*. R. Oldenbourg Verlag, München.

Strassen, V. (1965). The existence of probability measures with given marginals. *Annals of Mathematical Statistics*, **36**, 423–439.

Szekli, R. (1995). *Stochastic Ordering and Dependence in Applied Probability*. Volume 97 of *Lecture Notes in Statistics*, Springer-Verlag, New York.

Tchen, A. H. (1980). Inequalities for distributions with given marginals. *Annals of Probability*, **8**, 814–827.

Tong, Y. L. (1980). *Probability Inequalities in Multivariate Distributions*. Academic Press, Boston, Massachusetts.

Veinott, Jr., A. F. (1965). Optimal policy in a dynamic, single product, nonstationary inventory model with several demand classes. *Operations Research*, **13**, 761–778.

Chapter 3

The *q*-Factorial Moments of Discrete *q*-Distributions and a Characterization of the Euler Distribution

CH. A. CHARALAMBIDES AND N. PAPADATOS
Department of Mathematics, University of Athens,
Athens, Greece

CONTENTS

ABSTRACT

The classical discrete distributions binomial, geometric, and negative binomial are defined on the stochastic model of a sequence of independent and identical Bernoulli trials. The Poisson distribution may be defined as an approximation of the binomial (or negative binomial) distribution.

The corresponding q-distributions are defined on the more general stochastic model of a sequence of Bernoulli trials with probability of success at any trial depending on the number of trials. In this paper targeting the problem of calculating the moments of q-distributions, we introduce and study q-factorial moments, the calculation of which is as easy as the calculation of the factorial moments of the classical distributions. The usual factorial moments are connected with the q-factorial moments through the q-Stirling numbers of the first kind. Several examples, illustrating the method, are presented. Further, the Euler distribution is characterized through its q-factorial moments.

KEYWORDS AND PHRASES: q-distributions, q-moments, q-Stirling numbers

3.1   INTRODUCTION

Consider a sequence of independent Bernoulli trials with probability of success at the ith trial p_i, $i = 1, 2, \ldots$. The study of the distribution of the number X_n of successes up to the nth trial, as well as the closely related distribution of the number Y_k of trials until the occurrence of the kth success, has attracted special attention. In the particular case $p_i = \theta q^{i-1}/(1 + \theta q^{i-1})$, $i = 1, 2, \ldots$, $0 < q < 1$, $\theta > 0$, the distribution of the random variable X_n, called the q-binomial distribution, has been studied by Kemp and Newton (1990) and Kemp and Kemp (1991). The q-binomial distribution, for $n \to \infty$, converges to a q-analog of the Poisson distribution, called the Heine distribution. This distribution was introduced and examined by Benkherouf and Bather (1988). Kemp (1992a,b) further studied the Heine distribution. In the case $p_i = 1 - \theta q^{i-1}$, $i = 1, 2, \ldots$, $0 < q < 1$, $0 < \theta < 1$, the distribution of the random variable Y_k is called the q-Pascal distribution. A stochastic model described by Dunkl (1981) led to the particular case $\theta = q^{m-k+1}$ of this distribution. This distribution, also studied by Kemp (1998), is called the absorption distribution. For $k \to \infty$, the distribution of the number of failures until the occurrence of the kth success $W_k = Y_k - k$ converges to another q-analog of the Poisson distribution, called Euler the distribution. This distribution was studied by Benkherouf and Bather (1988) and Kemp (1992a,b). Kemp (2001) characterized the absorption distribution as the conditional distribution of a q-binomial distribution given the sum of a q-binomial and a Heine distribution with the same argument parameter.

In the present paper, we propose the introduction of q-factorial moments for q-distributions. These moments, apart from the interest in their own, may be used as an intermediate step in the evaluation of the usual moments of the q-distributions. In this respect, an expression of the usual factorial moments in terms of the q-factorial moments is derived. Several examples illustrating the method are presented and a characterization of the Euler distribution through its q-factorial moments is derived.

3.2 *q*-NUMBERS, *q*-FACTORIALS, AND *q*-STIRLING NUMBERS

Let $0 < q < 1$, x be a real number and k be a positive integer. The number $[x]_q = (1 - q^x)/(1 - q)$ is called a *q-real number*. In particular, $[k]_q$ is called a q-positive integer. The factorial of the q-number $[x]_q$ of order k, which is defined by

$$[x]_{k,q} = [x]_q [x - 1]_q \cdots [x - k + 1]_q$$
$$= \frac{(1 - q^x)(1 - q^{x-1}) \cdots (1 - q^{x-k+1})}{(1 - q)^k},$$

is called a *q-factorial of x of order k*. In particular, $[k]_q! = [1]_q [2]_q \cdots [k]_q$ is called a *q-factorial of k*. The *q-binomial coefficient* is defined by

$$\begin{bmatrix} x \\ k \end{bmatrix}_q = \frac{[x]_{k,q}}{[k]_q!} = \frac{(1 - q^x)(1 - q^{x-1}) \cdots (1 - q^{x-k+1})}{(1 - q)(1 - q^2) \cdots (1 - q^k)}.$$

Note that

$$\lim_{q \to 1} \begin{bmatrix} x \\ k \end{bmatrix}_q = \begin{pmatrix} x \\ k \end{pmatrix}.$$

The *q-binomial* and the *negative q-binomial* expansions are expressed as

$$\prod_{i=1}^{n}(1 + tq^{i-1}) = \sum_{k=0}^{n} q^{k(k-1)/2} \begin{bmatrix} n \\ k \end{bmatrix}_q t^k, \tag{3.2.1}$$

and

$$\prod_{i=1}^{n}(1 - tq^{i-1})^{-1} = \sum_{k=0}^{\infty} \begin{bmatrix} n+k-1 \\ k \end{bmatrix}_q t^k, \quad |t| < 1, \tag{3.2.2}$$

respectively. In general, the transition of an expression to a *q-analog* is not unique. Other *q-binomial* and *negative q-binomial expansions* useful in the sequel are the following:

$$(1 - (1-q)[t]_q)^n = (q^t)^n$$

$$= \sum_{k=0}^{n}(-1)^k q^{k(k-1)/2}(1-q)^k \begin{bmatrix} n \\ k \end{bmatrix}_q [t]_{k,q} \tag{3.2.3}$$

and

$$(1 - (1-q)[t]_q)^{-n} = (q^t)^{-n}$$

$$= \sum_{k=0}^{\infty} q^{-nk}(1-q)^k \begin{bmatrix} n+k-1 \\ k \end{bmatrix}_q [t]_{k,q}. \tag{3.2.4}$$

Also useful are the following two *q-exponential functions*:

$$e_q(t) = \prod_{i=1}^{\infty}(1 - (1-q)q^{i-1}t)^{-1}$$

$$= \sum_{k=0}^{\infty} \frac{t^k}{[k]_q!}, \quad |t| < 1/(1-q), \tag{3.2.5}$$

$$E_q(t) = \prod_{i=1}^{\infty}(1 + (1-q)q^{i-1}t)$$

$$= \sum_{k=0}^{\infty} q^{k(k-1)/2} \frac{t^k}{[k]_q!}, \quad -\infty < t < \infty, \tag{3.2.6}$$

with $e_q(t)E_q(-t) = 1$. The nth-order q-factorial $[t]_{n,q}$ is expanded into powers of the q-number $[t]_q$ and inversely as follows:

$$[t]_{n,q} = q^{-n(n-1)/2} \sum_{k=0}^{n} s_q(n,k)[t]_q^k, \quad n = 0, 1, \ldots, \qquad (3.2.7)$$

$$[t]_q^n = \sum_{k=0}^{n} q^{k(k-1)/2} S_q(n,k)[t]_{k,q}, \quad n = 0, 1, \ldots. \qquad (3.2.8)$$

The coefficients $s_q(n,k)$ and $S_q(n,k)$ are called *q-Stirling numbers of the first and second kind*, respectively. Closed expressions, recurrence relations, and other properties of these numbers are examined by Carlitz (1933, 1948) and Gould (1961).

3.3 *q*-FACTORIAL MOMENTS

The calculation of the mean and the variance and generally the calculation of the moments of a discrete q-distribution are quite difficult. Several techniques have been used for the calculation of the mean and the variance of particular q-distributions. The general method of evaluation of moments by differentiating the probability generating function, used by Kemp (1992a, 1998), is bounded to the calculation of the first two moments. This limited applicability is due to the inherent difficulties in the differentiation of the hypergeometric series. We propose the introduction of the q-factorial moments of q-distributions, the calculation of which is as easy as that of the usual factorial moments of the classical discrete distributions.

DEFINITION 3.3.1 Let X be a nonnegative integer valued random variable with probability mass function $f(x) = P(X = x)$, $x = 0, 1, \ldots$.

(a) The mean of the rth-order q-factorial $[X]_{r,q}$,

$$E([X]_{r,q}) = \sum_{x=r}^{\infty} [x]_{r,q} f(x), \qquad (3.3.1)$$

provided it exists, is called the rth-order (descending) q-factorial moment of the random variable X.

(b) The mean of the rth-order ascending q-factorial $[X + r - 1]_{r,q}$,

$$E([X + r - 1]_{r,q}) = \sum_{x=1}^{\infty} [x + r - 1]_{r,q} f(x), \qquad (3.3.2)$$

provided it exists, is called the rth-order ascending q-factorial moment of the random variable X.

The usual factorial moments are expressed in terms of the q-factorial moments, through the q-Stirling number of the first kind, in the following theorem.

THEOREM 3.3.1 *Let $E([X]_{r,q})$ and $E([X + r - 1]_{r,q})$ be the rth-order descending and ascending q-factorial moments, $r = 1, 2, \ldots$, respectively, of a nonnegative integer valued random variable X. Then*

$$E[(X)_m] = m! \sum_{r=m}^{\infty} (-1)^{r-m} s_q(r, m)$$

$$\times \frac{(1-q)^{r-m}}{[r]_q!} E([X]_{r,q}), \qquad (3.3.3)$$

and

$$E[(X + m - 1)_m]$$

$$= m! \sum_{r=m}^{\infty} q^{-\binom{r}{2}} s_q(r, m)$$

$$\times \frac{(1-q)^{r-m}}{[r]_q!} E(q^{-rX} [X + r - 1]_{r,q}), \qquad (3.3.4)$$

provided the series are convergent. The coefficient $s_q(r, k)$ is the q-Stirling number of the first kind.

PROOF. According to Newton's binomial formula, for x nonnegative integer, we have

$$(1 - (1-q)[t]_q)^x = \sum_{k=0}^{x} (-1)^k \binom{x}{k} (1-q)^k [t]_q^k$$

while, from (3.2.3) and (3.2.7) we get

$$(1 - (1 - q)[t]_q)^x = \sum_{k=0}^{x} \left\{ \sum_{r=k}^{x} (-1)^r (1 - q)^r s_q(r, k) \begin{bmatrix} x \\ r \end{bmatrix}_q \right\} [t]_q^k$$

so

$$\binom{x}{k} = \sum_{r=k}^{x} (-1)^{r-k} (1 - q)^{r-k} s_q(r, k) \begin{bmatrix} x \\ r \end{bmatrix}_q .$$

Multiplying both members of this expression by the probability mass function $f(x)$ of the random variable X and summing for all $x = 0, 1, \ldots$, we deduce, according to (3.3.1), the required expression (3.3.3).

Similarly, expanding both members of (3.2.4) into powers of $[t]_q$ with the aid of Newton's negative binomial formula and expression (3.2.7) and taking expectations in the resulting expression, (3.3.4) is deduced. ∎

Note that Dunkl (1981), starting from Newton's polynomial expression of a function in terms of divided differences at certain points and letting the function be the binomial coefficient $\binom{x}{k}$ and the points be the q-numbers $[r]_q, r = 0, 1, \ldots, x$, first derived expression (3.3.3).

In the following examples the q-factorial moments and the usual factorial moments of several discrete q-distributions are evaluated.

EXAMPLE 3.3.1 *q-binomial distribution.* Consider a sequence of independent Bernoulli trials with probability of success at the ith trial $p_i = \theta q^{i-1}/(1 + \theta q^{i-1})$, $i = 1, 2, \ldots$, $0 < q < 1$, $\theta > 0$. The probability mass function of the number X_n of successes up to the nth trial is given by

$$f_{X_n}(x) = \begin{bmatrix} n \\ x \end{bmatrix}_q q^{x(x-1)/2} \theta^x \prod_{i=1}^{n} (1 + \theta q^{i-1})^{-1}, \quad x = 0, 1, \ldots, n,$$

with $0 < q < 1$, $\theta > 0$. The rth-order q-factorial moment of the random variable X_n, according to Definition 3.3.1, is given by the sum

$$E([X_n]_{r,q}) = \frac{1}{\prod_{i=1}^{n} (1 + \theta q^{i-1})} \sum_{x=r}^{n} [x]_{r,q} \begin{bmatrix} n \\ x \end{bmatrix}_q q^{x(x-1)/2} \theta^x$$

and since

$$[x]_{r,q} \begin{bmatrix} n \\ x \end{bmatrix}_q = [n]_{r,q} \begin{bmatrix} n-r \\ x-r \end{bmatrix}_q,$$

$$\binom{x}{2} = \binom{x-r}{2} + \binom{r}{2} + r(x-r),$$

it is written as

$$E([X]_n]_{r,q}) = \frac{[n]_{r,q} \, q^{r(r-1)/2}\theta^r}{\prod_{i=1}^{n}(1+\theta q^{i-1})}$$

$$\times \sum_{x=r}^{n} q^{(x-r)(x-r-1)/2} \begin{bmatrix} n-r \\ x-r \end{bmatrix}_q (\theta q^r)^{x-r}$$

and, by the q-binomial formula (3.2.1), reduces to

$$E([X_n]_{r,q}) = \frac{[n]_{r,q} \, q^{r(r-1)/2}\theta^r}{\prod_{i=1}^{r}(1+\theta q^{i-1})}.$$

The kth-order factorial moment of the random variable X_n, according to Theorem 3.3.1, is given by

$$E[(X_n)_k] = k! \sum_{r=k}^{n}(-1)^{r-k}s_q(r,k)\frac{(1-q)^{r-k}q^{r(r-1)/2}\theta^r}{\prod_{i=1}^{r}(1+\theta q^{i-1})} \begin{bmatrix} n \\ r \end{bmatrix}_q.$$

EXAMPLE 3.3.2 *Heine distribution.* The probability mass function of the q-binomial distribution for $n \to \infty$ converges to the probability mass function of the Heine distribution

$$f_X(x) = e_q(-\lambda)\frac{q^{x(x-1)/2}\lambda^x}{[x]_q!}, \quad x = 0,1,\ldots,$$

with $0 < q < 1$, $\lambda > 0$, where $\lambda = \theta/(1-q)$ and $e_q(-\lambda) = \prod_{i=1}^{\infty} (1+\lambda(1-q)q^{i-1})^{-1}$ is the q-exponential function (3.2.5). The rth-order q-factorial moment of the random variable X is given by

$$E([X]_{r,q}) = e_q(-\lambda) \sum_{x=r}^{\infty}[x]_{r,q}\frac{q^{x(x-1)/2}\lambda^x}{[x]_q!}$$

$$= q^{r(r-1)/2}\lambda^r e_q(-\lambda) \sum_{x=r}^{\infty} \frac{q^{(x-r)(x-r-1)/2}(\lambda q^r)^{x-r}}{[x-r]_q!}$$

and since

$$\sum_{x=r}^{\infty} \frac{q^{(x-r)(x-r-1)/2}(\lambda q^r)^{x-r}}{[x-r]_q!} = E_q(\lambda q^r)$$

$$= \prod_{i=1}^{\infty} (1 + \lambda(1-q)q^{r+i-1})$$

it reduces to

$$E([X]_{r,q}) = \frac{q^{r(r-1)/2}\lambda^r}{\prod_{i=1}^{r}(1 + \lambda(1-q)q^{i-1})}.$$

Further, by Theorem 3.3.1,

$$E[(X)_k] = k! \sum_{r=k}^{\infty} (-1)^{r-k} s_q(r,k)$$

$$\times \frac{(1-q)^{r-k}q^{r(r-1)/2}}{\prod_{i=1}^{r}(1 + \lambda(1-q)q^{i-1})} \cdot \frac{\lambda^r}{[r]_q!}.$$

EXAMPLE 3.3.3 *q-Pascal distribution.* Consider a sequence of independent Bernoulli trials with probability of success at the ith trial $p_i = 1 - \theta q^{i-1}$, $i = 1, 2, \ldots$, $0 < q < 1$, $0 < \theta < 1$. The probability mass function of the number Y_k of trials until the occurrence of the kth success is given by

$$f_{Y_k}(y) = \begin{bmatrix} y-1 \\ k-1 \end{bmatrix}_q \theta^{y-k} \prod_{i=1}^{k}(1 - \theta q^{i-1}), \quad y = k, k+1, \ldots,$$

with $0 < q < 1$, $0 < \theta < 1$. The rth-order ascending q-factorial moment of the random variable Y_k, according to Definition 3.3.1, is given by the sum

$$E([Y_k + r - 1]_{r,q}) = \frac{1}{\prod_{i=1}^{k}(1 - \theta q^{i-1})-1}$$

$$\times \sum_{y=k}^{\infty} [y + r - 1]_{r,q} \begin{bmatrix} y-1 \\ k-1 \end{bmatrix}_q \theta^{y-k}$$

and since

$$[y + r - 1]_{r,q} \begin{bmatrix} y-1 \\ k-1 \end{bmatrix}_q = [k + r - 1]_{r,q} \begin{bmatrix} y+r-1 \\ k+r-1 \end{bmatrix}_q,$$

it is written as

$$E([Y_k + r - 1]_{r,q}) = \frac{[k + r - 1]_{r,q}}{\prod_{i=1}^{k}(1 - \theta q^{i-1})^{-1}}$$

$$\times \sum_{y=k}^{\infty} \begin{bmatrix} y + r - 1 \\ k + r - 1 \end{bmatrix}_q \theta^{y-k}.$$

Thus, by the q-negative binomial formula (3.2.2),

$$E([Y_k + r - 1]_{r,q}) = \frac{[k + r - 1]_{r,q}}{\prod_{i=1}^{r}(1 - \theta q^{k+i-1})}.$$

Similarly,

$$E(q^{-rY_k}[Y_k + r - 1]_{r,q}) = \frac{[k + r - 1]_{r,q} q^{-kr}}{\prod_{i=1}^{r}(1 - \theta q^{-r+i-1})}$$

and by Theorem 3.3.1,

$$E[(X + m - 1)_m] = m! \sum_{r=m}^{\infty} q^{-kr - \binom{r}{2}} s_q(r, m)$$

$$\times \frac{(1 - q)^{r-m}}{\prod_{i=1}^{r}(1 - \theta q^{-r+i-1})} \begin{bmatrix} k + r - 1 \\ r \end{bmatrix}_q.$$

EXAMPLE 3.3.4 *Euler distribution.* The probability mass function of the number of failures until the occurrence of the kth success $W_k = Y_k - k$ is given by

$$f_{W_k}(w) = \begin{bmatrix} k + w - 1 \\ w \end{bmatrix}_q \theta^w \prod_{i=1}^{k}(1 - \theta q^{i-1}), \quad w = 0, 1, \dots.$$

This distribution, which may be called the q-negative binomial distribution, for $k \to \infty$, converges to the Euler distribution with probability mass function

$$f_X(x) = E_q(-\lambda)\frac{\lambda^x}{[x]_q!}, \quad x = 0, 1, \dots,$$

with $0 < q < 1$, $0 < \lambda < 1/(1 - q)$, where $\lambda = \theta/(1 - q)$ and $E_q(-\lambda) = \prod_{i=1}^{\infty}(1 - \lambda(1 - q)q^{i-1})$ is the q-exponential function (3.2.6). The rth-order q-factorial moment of the random variable X is given by

$$E([X]_{r,q}) = E_q(-\lambda) \sum_{x=r}^{\infty} [x]_{r,q} \frac{\lambda^x}{[x]_q!} = \lambda^r E_q(-\lambda) \sum_{x=r}^{\infty} \frac{\lambda^{x-r}}{[x-r]_q!}$$

and since, by (3.2.5),

$$\sum_{x=r}^{\infty} \frac{\lambda^{x-r}}{[x-r]_q!} = e_q(\lambda) = \frac{1}{E_q(-\lambda)},$$

it follows that

$$E([X]_{r,q}) = \lambda^r.$$

Further, by Theorem 3.3.1,

$$E[(X)_k] = k! \sum_{r=k}^{\infty} (-1)^{r-k} s_q(r,k) \frac{(1-q)^{r-k}\lambda^r}{[r]_q!}.$$

3.4 A CHARACTERIZATION OF THE EULER DISTRIBUTION

Consider a family of nonnegative integer valued random variables $\{X_\lambda, 0 < \lambda < \rho \leq \infty\}$ having a power series distribution with probability mass function

$$f(x; \lambda) = \frac{a(x)\lambda^x}{g(\lambda)}, \quad x = 0, 1, \ldots, \quad 0 < \lambda < \rho \qquad (3.4.1)$$

and series function

$$g(\lambda) = \sum_{x=0}^{\infty} a(x)\lambda^x, \quad 0 < \lambda < \rho.$$

It is well known that the mean-variance equality

$$E(X_\lambda) = Var(X_\lambda) \quad \text{for all} \quad \lambda \in (0, \rho)$$

characterizes the Poisson family of distributions [see Kosambi (1949) and Patil (1962)]. Note that the requirement that this equality holds for all $\lambda \in (0, \rho)$ has been overlooked by some authors [see Sapatinas (1994) for details]. This requirement may be relaxed by weaker ones; e.g., it suffices to verify it for all $\lambda \in I$, where I is any nondegenerate subinterval of $(0, \rho)$. A q-analogue to the Kosambi–Patil characterization for the Euler distribution is derived in the following theorem.

THEOREM 3.4.1 *Assume that a family of nonnegative integer valued random variables* $\{X_\lambda, 0 < \lambda < \rho \leq \infty\}$ *obeys a power series distribution with probability mass function* (3.4.1). *Then,* X_λ *has an Euler distribution if and only if*

$$E([X_\lambda]_{2,q}) = [E([X_\lambda]_q)]^2 \qquad (3.4.2)$$

for all $\lambda \in (0, \rho)$.

PROOF. Assume first that (3.4.2) holds for all $\lambda \in (0, \rho)$. Then

$$\frac{\lambda^2}{g(\lambda)} \sum_{x=0}^{\infty} [x+1]_q [x+2]_q \, a(x+2)\lambda^x$$

$$= \frac{\lambda^2}{[g(\lambda)]^2} \left[\sum_{x=0}^{\infty} [x+1]_q \, a(x+1)\lambda^x \right]^2$$

which, using the series $g(\lambda) = \sum_{x=0}^{\infty} a(x)\lambda^x$, may be written as

$$\left[\sum_{x=0}^{\infty} a(x)\lambda^x \right] \left[\sum_{x=0}^{\infty} [x+1]_q [x+2]_q \, a(x+2)\lambda^x \right]$$

$$= \left[\sum_{x=0}^{\infty} [x+1]_q \, a(x+1)\lambda^x \right]^2$$

or equivalently as

$$\sum_{x=0}^{\infty} \left\{ \sum_{k=0}^{x} [k+1]_q [k+2]_q \, a(k+2)a(x-k) \right\} \lambda^x$$

$$= \sum_{x=0}^{\infty} \left\{ \sum_{k=0}^{x} [k+1]_q [x-k+1]_q \, a(k+1)a(x-k+1) \right\} \lambda^x.$$

Hence,

$$\sum_{k=0}^{x} [k+1]_q [k+2]_q \, a(k+2)a(x-k)$$

$$= \sum_{k=0}^{x} [k+1]_q [x-k+1]_q \, a(k+1)a(x-k+1), \qquad (3.4.3)$$

for $x = 0, 1, \dots$. Setting $x = 0$, it follows that

$$[2]_q! \, a(2)a(0) = [a(1)]^2.$$

It is easy to see that if $a(0) = 0$, then $a(1) = 0$ and, using (3.4.3), it follows that $a(x) = 0$ for all $x = 0, 1, \ldots$, which is a contradiction to the assumption that X_λ obeys a power series distribution. Thus, $a(0) \neq 0$, and without any loss of generality we may assume that $a(0) = 1$. Therefore,

$$a(2) = a^2/[2]_q!,$$

where $a = a(1) > 0$. Further, setting $x = 1$ it follows that

$$[2]_q! a(2)a(1) + [3]_q! a(3)a(0) = 2[2]_q! a(2)a(1)$$

and

$$a(3) = a^3/[3]_q!.$$

Suppose that

$$a(k) = a^k/[k]_q!, \quad k = 0, 1, \ldots, x+1.$$

Then

$$[x+1]_q[x+2]_q \, a(x+2)a(0)$$

$$+ \sum_{k=0}^{x-1} [k+1]_q[k+2]_q \, a(k+2)a(x-k)$$

$$= [x+1]_q[1]_q \, a(x+1)a(1)$$

$$+ \sum_{k=0}^{x-1} [k+1]_q[x-k+1]_q \, a(k+1)a(x-k+1)$$

and since

$$\sum_{k=0}^{x-1} [k+1]_q[k+2]_q \, a(k+2)a(x-k)$$

$$= a^{x+2} \sum_{k=0}^{x-1} \frac{1}{[k]_q![x-k]_q!},$$

$$\sum_{k=0}^{x-1} [k+1]_q[x-k+1]_q \, a(k+1)a(x-k+1)$$

$$= a^{x+2} \sum_{k=0}^{x-1} \frac{1}{[k]_q![x-k]_q!},$$

it follows that

$$[x + 2]_q \, a(x + 2) = a \cdot a(x + 1)$$

so

$$a(x + 2) = a^{x+2}/[x + 2]_q!.$$

Thus,

$$a(x) = a^x/[x]_q!, \quad x = 0, 1, \ldots.$$

Further, the series function, by (3.2.5), is given by

$$g(\lambda) = \sum_{x=0}^{\infty} a(x)\lambda^x = \sum_{x=0}^{\infty} \frac{(a\lambda)^x}{[x]_q!} = e_q(a\lambda) = \frac{1}{E_q(-a\lambda)}$$

for $0 < q < 1$, $0 < a\lambda < 1/(1 - q)$, so the random variable X_λ has an Euler distribution with probability mass function

$$f(x; \lambda) = E_q(-a\lambda)\frac{(a\lambda)^x}{[x]_q!}, \quad x = 0, 1, \ldots,$$

with $0 < q < 1$, $0 < a\lambda < 1/(1 - q)$. Finally, according to Example 3.3.4, the q-factorial moments of the Euler distribution satisfy relation (3.4.2). ∎

ACKNOWLEDGMENTS

This research was partially supported by the University of Athens Research Special Account under grants 70/4/3406 and 70/4/5637.

REFERENCES

Benkherouf, L. and Bather, J. A. (1988). Oil exploration: sequential decisions in the face of uncertainty. *Journal of Applied Probability*, **25**, 529–543.

Blomqvist, N. (1952). On an exhaustion process. *Skandinavisk Akktuarietidskrift*, **36**, 201–210.

Carlitz, L. (1933). On Abelian fields. *Transactions of the American Mathematical Society*, **35**, 122–136.

Carlitz, L. (1948). q-Bernoulli numbers and polynomials. *Duke Mathematical Journal*, **15**, 987–1000.

Dunkl, C. F. (1981). The absorption distribution and the q-binomial theorem. *Communications in Statistics—Theory and Methods*, **10**, 1915–1920.

Gould, H. W. (1961). The q-Stirling numbers of the first and second kinds. *Duke Mathematical Journal*, **28**, 281–289.

Kemp, A. (1998). Absorption sampling and the absorption distribution. *Journal of Applied Probability*, **35**, 489–494.

Kemp, A. (2001). A characterization of a distribution arising from absorption sampling. In *Probability and Statistical Models with Applications* (Eds., Ch. A. Charalambides, M. V. Koutras, and N. Balakrishnan), pp. 239–246, Chapman & Hall/CRC Press, Boca Raton, Florida.

Kemp, A. (1992a). Heine–Euler extensions of Poisson distribution. *Communications in Statistics—Theory and Methods*, **21**, 791–798.

Kemp, A. (1992b). Steady-state Markov chain models for Heine and Euler distributions. *Journal of Applied Probability*, **29**, 869–876.

Kemp, A. and Kemp, C. D. (1991). Weldon's dice date revisited. *The American Statistician*, **45**, 216–222.

Kemp, A. and Newton, J. (1990). Certain state-dependent processes for dichotomized parasite populations. *Journal of Applied Probability*, **27**, 251–258.

Kosambi, D. D. (1949). Characteristic properties of series distributions. *Proceedings of the National Institute for Science, India*, **15**, 109–113.

Patil, G. P. (1962). Certain properties of the generalized power series distribution. *Annals of the Institute of Statistical Mathematics*, **14**, 179–182.

Sapatinas, T. (1994). Letter to the editor. A remark on P. C. Consul's (1990) counterexample: "New class of location-parameter discrete probability distributions and their characterizations." [*Communications in Statistics—Theory and Methods*, **19** (1990), 4653–4666] with a rejoinder by Consul. *Communications in Statistics—Theory and Methods*, **23**, 2127–2130.

Chapter 4

On the Characterization of Distributions through the Properties of Conditional Expectations of Order Statistics

I. BAIRAMOV
Department of Mathematics, İzmir University of Economics, İzmir, Turkey

O. GEBIZLIOGLU
Department of Statistics, Ankara University, Ankara, Turkey

CONTENTS

ABSTRACT

We present results that lead to characterization of continuous distributions using conditional expectations of order statistics.

KEYWORDS AND PHRASES: Characterization, conditional expectations, truncated data, order statistics

4.1   INTRODUCTION

Characterization of distributions through the properties of conditional expectations of order statistics and record values aroused interest of many authors. Let $X_1, X_2, \ldots, X_n, \ldots$ be a sequence of iid random variables with cdf F. Denote $X_{1:n}$, $X_{2:n}, \ldots, X_{n:n}$ the order statistics of $X_1, X_2, \ldots, X_n$. Ferguson (1967) considered the problem to determine all df's for which $E\{X_{k+m:n} \mid X_{k:n}\} = aX_{k:n} + b$ (a.s.) when F is a continuous distribution function. This problem was developed by Nagaraja (1988a,b) (who considered the discrete distributions also) and by many authors including Beg and Kirmani (1974), Dallas (1973), Wang and Srivastava (1980), and Beg and Balasubramanian (1990). In more general form, the problem of characterizing of distribution by the regression of order statistics and record values has been considered by Wesolowski and Ahsanullah (1997, 1998), Blazquez and Rebollo (1997), and Dembinska and Wesolowski (2000). Beg and Balasubramanian (1990) characterize the distributions for which the explicit form of the distribution function is known, continuous, and strictly increasing in its support (a_X, b_X) through the property,

$$E\left\{\frac{1}{s-1}\sum g(X_{i:n}) \mid X_{s:n} = x\right\}$$

$$= \frac{g(x) + g(a_X+)}{2}, \quad \forall x \in (a_X, b_X)$$

by a suitable choice of nonconstant continuous function g. Balasubramanian and Dey (1997) have proved that if X is a random variable with absolutely continuous cdf F, then

$$E\left\{\sum_{i=r+1}^{s-1}\frac{g(X_{i:n})}{s-r-1} \mid X_{r:n}, X_{s:n}\right\} = \frac{2g(X_{r:n})g(X_{s:n})}{g(X_{r:n}) + g(X_{s:n})}$$

if and only if $g(x) = \sqrt{\frac{b}{F(x)+a}}$. An application of this theorem, for example, gives the following interesting characterization. Distributions with cdf of the form $F(x) = Ax^k + B$, which include power function distributions, Pareto distribution, and Rectangular distribution, are characterized by any one of the following two conditions:

$$E\left\{\frac{1}{s-r-1}\sum_{i=r+1}^{s-1}\frac{1}{X_{i:n}^{k/2}}\mid X_{r:n}, X_{s:n}\right\} = \frac{2}{X_{r:n}^{k/2} + X_{s:n}^{k/2}};$$

$$E\left\{\frac{1}{s-r-1}\sum_{i=r+1}^{s-1}\frac{1}{X_{i:n}^{2k}}\mid X_{r:n}, X_{s:n}\right\} = \frac{1}{X_{r:n}^{k} + X_{s:n}^{k}}.$$

Recently Bairamov and Apaydin (2000) have prove that if $h(x)$ is a differentiable real valued function, such that for all $x \in (0, 1)$

$$h'(x) \neq \frac{h(1) - h(x)}{1 - x},$$

then X has the distribution function F if and only if the representation

$$E\left\{\frac{1}{n-r}\sum_{i=r+1}^{n}g(X_{i:n})\mid X_{r:n} = y\right\} = \frac{h(1) - h(F(y))}{1 - F(y)}$$

holds for all $a_X < y < b_X$, where $g(x) = h'(F(x))$.

In this paper we extend this result and give characterizations for continuous distributions using expectations of left and right censored samples.

4.2 CHARACTERIZATION USING LEFT AND RIGHT TRUNCATION

Let X be a continuous random variable with the distribution function F. We denote by $F^{-1}(y) = \inf\{x : F(x) \geq y\}$, $0 < y < 1$, the inverse distribution function and by $a_X = \inf\{x : F(x) > 0\}$ and $b_X = \inf\{x : F(x) < 1\}$ the endpoints of support of X.

THEOREM 4.2.1 *Let $h(x)$ be a continuous real valued function such that $h(x) < \infty$ and the condition*

$$h'(y) \neq \frac{h(y) - h(x)}{y - x}$$

is valid for all $0 < x < y < 1$. Then a continuous random variable X has distribution function G if and only if the representation

$$E\{h'(G(X)) \mid x \leq X \leq y\} = \frac{h(G(y)) - h(G(x))}{G(y) - G(x)} \qquad (4.2.1)$$

is valid for all $a_X < x < y < b_X$.

PROOF. It is clear that (4.2.1) is equivalent to

$$E\{h'(G(X)) \mid x \leq G(X) \leq y\} = \frac{h(y) - h(x)}{y - x}$$

for all $0 < x < y < 1$. To prove *necessity*, let X have df $G(x)$. Denote $G'(x) = g(x)$. Then

$$E\{h'(G(X)) \mid x \leq X \leq y\} = \frac{1}{G(y) - G(x)} \int_x^y h'(G(z)) g(z) \, dz$$

$$= \frac{h(G(y)) - h(G(x))}{G(y) - G(x)}.$$

Sufficiency. Let (4.2.1) hold. Denote by $F(x)$ and $f(x)$ the df and pdf of X, respectively. Then one has from (4.2.1)

$$\frac{1}{F(y) - F(x)} \int_x^y h'(G(z)) f(z) \, dz = \frac{h(G(y)) - h(G(x))}{G(y) - G(x)}$$

or

$$\int_x^y h'(G(z)) f(z) \, dz = \frac{h(G(y)) - h(G(x))}{G(y) - G(x)} [F(y) - F(x)].$$

$$(4.2.2)$$

Differentiating (4.2.2) with respect to y, we have

$$h'(G(y))f(y)$$

$$= \frac{\{f(y)[G(y) - G(x)] - g(y)[F(y) - F(x)]\}}{[G(y) - G(x)]^2} [h(G(y)) - h(G(x))]$$

$$+ \frac{F(y) - F(x)}{[G(y) - G(x)]} h'(G(y))g(y)$$

$$= \frac{f(y)[G(y) - G(x)][h(G(y)) - h(G(x))]}{[G(y) - G(x)]^2}$$

$$- \frac{g(y)[F(y) - F(x)][h(G(y)) - h(G(x))]}{[G(y) - G(x)]^2}$$

$$+ \frac{F(y) - F(x)}{G(y) - G(x)} h'(G(y))g(y).$$

Hence,

$$h'(G(y))f(y) = \frac{f(y)[h(G(y)) - h(G(x))]}{G(y) - G(x)}$$

$$- \frac{g(y)[F(y) - F(x)][h(G(y)) - h(G(x))]}{[G(y) - G(x)]^2}$$

$$+ \frac{F(y) - F(x)}{G(y) - G(x)} h'(G(y))g(y). \qquad (4.2.3)$$

Dividing (4.2.3) by $F(y) - F(x)$, we have

$$h'(G(y)) \frac{f(y)}{F(y) - F(x)} = \frac{f(y)}{F(y) - F(x)} \frac{h(G(y)) - h(G(x))}{G(y) - G(x)}$$

$$- \frac{g(y)[h(G(y)) - h(G(x))]}{[G(y) - G(x)]^2}$$

$$+ \frac{h'(G(y))g(y)}{G(y) - G(x)}. \qquad (4.2.4)$$

From (4.2.4),

$$h'(G(y)) \left[\frac{f(y)}{F(y) - F(x)} - \frac{g(y)}{G(y) - G(x)} \right]$$
$$= [h(G(y)) - h(G(x))]$$
$$\times \left[\frac{f(y)}{[F(y) - F(x)][G(y) - G(x)]} - \frac{g(y)}{[G(y) - G(x)]^2} \right].$$

(4.2.5)

From (4.2.5) we have

$$h'(G(y)) \frac{1}{[F(y) - F(x)][G(y) - G(x)]}$$
$$\times \{f(y)[G(y) - G(x)] - g(y)[F(y) - F(x)]\}$$
$$= \frac{h(G(y)) - h(G(x))}{[F(y) - F(x)][G(y) - G(x)]^2}$$
$$\times [f(y)[G(y) - G(x)] - g(y)[F(y) - F(x)]]$$

and

$$\{f(y)[G(y) - G(x)] - g(y)[F(y) - F(x)]\}$$
$$\times \left\{ h'(G(y)) - \frac{h(G(y)) - h(G(x))}{G(y) - G(x)} \right\} = 0.$$

Therefore,

$$\frac{f(y)}{F(y) - F(x)} = \frac{g(y)}{G(y) - G(x)} \forall x < y.$$

Taking $x = b_X$, we have $\frac{f(y)}{F(y)} = \frac{g(y)}{G(y)}$ for all $a_X < y < b_X$. The theorem is thus proved. ∎

LEMMA 4.2.1 *Let $X_1, X_2, \ldots, X_n$ be a sample of size n from a population with distribution function F. Then for $1 \leq r < s \leq n$, the conditional joint pdf of $X_{r+1:n}, X_{r+2:n}, \ldots, X_{s:n}$ with the condition that $X_{r:n} = x$ and $X_{r:n} = y$ coincide with the joint pdf of order statistics $Y_{1:s-r-1} \leq Y_{2:s-r-1} \leq \cdots \leq Y_{s-r-1:s-r-1}$ constructed from the sample $Y_1, Y_2, \ldots, Y_{s-r-1}$ of size $s - r - 1$ obtained from restricted random variable $Y \equiv X \mid x \leq X \leq y$.*

THEOREM 4.2.2 *Under conditions of Theorem 4.2.1, X has distribution function $G(x)$ if and only if the representation*

$$E\left\{\frac{1}{s-r-1}\sum_{i=r+1}^{s-1} g(X_{i:n}) \mid X_{r:n} = x, X_{s:n} = y\right\}$$
$$= \frac{h(G(y)) - h(G(x))}{G(y) - G(x)}$$

holds for all $a_X < x < y < b_X$, where $g(x) = h'(G(x))$.

PROOF. In fact, using Lemma 4.2.1 we have

$$E\left\{\frac{1}{s-r-1}\sum_{i=r+1}^{s-1} g(X_{i:n}) \mid X_{r:n} = x, X_{s:n} = y\right\}$$

$$= \frac{1}{s-r-1}\sum_{i=r+1}^{s-1} E\left\{g(X_{i:n}) \mid X_{r:n} = x, X_{s:n} = y\right\}$$

$$= \frac{1}{s-r-1}\sum_{i=1}^{s-r-1} E\left\{g(Y_{i:s-r-1})\right\}$$

$$= \frac{1}{s-r-1}\sum_{i=1}^{s-r-1} E\left\{g(Y_i)\right\}$$

$$= E\left\{g(Y_1)\right\} = E\left\{g(X) \mid x \le X \le y\right\}.$$

The proof is completed by using Theorem 4.2.1. ∎

4.3 CHARACTERIZATIONS FOR SOME IMPORTANT DISTRIBUTIONS

1. *Uniform distribution*

   (a)   The continuous random variable X has uniform distribution over $(0, 1)$ if and only if the representation

$$E\{X^k \mid x \le X \le y\} = \frac{y^{k+1} - x^{k+1}}{(k+1)(y-x)}$$

         holds for all $0 < x < y < 1$ and some integer k.

(b) The continuous random variable X has uniform distribution over $(0, 1)$ if and only if the representation

$$E\left\{\frac{1}{s-r-1}\sum_{i=r+1}^{s-1} X_{i:n}^k \mid X_{r:n} = x, X_{s:n} = y\right\}$$

$$= \frac{y^{k+1} - x^{k+1}}{(k+1)(y-x)}$$

holds for all $0 < x < y < 1$.

These results are obtained from Theorems 4.2.1 and 4.2.2 by using target function $h(x) = \frac{x^{k+1}}{k+1}, k \geq 1$.

2. *Weibull distribution*

(a) The continuous random variable X has Weibull distribution $F(x) = 1 - \exp(-\alpha x^\beta)$, $x \geq 0$, $\alpha > 0$, $\beta > 0$, if and only if the representation

$$E\{X^\beta \mid x \leq X \leq y\} = \frac{y^\beta - x^\beta}{\exp(-\alpha x^\beta) - \exp(\alpha x^\beta)}$$

holds for all $0 < x < y < \infty$.

(b) The continuous random variable X has Weibull distribution $F(x) = 1 - \exp(-\alpha x^\beta)$, $x \geq 0$, $\alpha > 0$, $\beta > 0$, if and only if the representation

$$E\left\{\frac{1}{s-r-1}\sum_{i=r+1}^{s-1} X_{i:n}^\beta \mid X_{r:n} = x, X_{s:n} = y\right\}$$

$$= \frac{y^\beta - x^\beta}{\exp(-\alpha x^\beta) - \exp(-\alpha x^\beta)}$$

holds for all $0 < x < y < \infty$, where $X_{i:n}$ is the order statistic of the sample of size n from X.

These results are obtained from Theorems 4.2.1 and 4.2.2 by using the target function $h(x) = (1 - x)(\ln(1 - x) - 1 + x)$ which has derivative $h'(x) = -\ln(1 - x)$.

REFERENCES

Balasubramanian, K. and Dey, A. (1997). Distributions characterized through conditional expectations. *Metrika*, **45**, 189–196.

Bairamov, I. G. and Apaidin, A. (2000). Characterization of continuous distributions by property of conditional expectation. *South African Statistical Journal*, **34**, 39–50.

Beg, M. I. and Balasubramanian, K. (1990). Distributions determined by conditioning on single order statistics. *Metrika*, **37**, 37–43.

Beg, M. I. and Kirmani, S. N. U. A. (1974). On a characterization of exponential and related distributions. *Australian Journal of Statistics*, **16**, 163–166.

Blazquez, F. L. and Rebollo, J. L. M. (1997). A characterization of distributions based on linear regression of order statistics and record values. *Sankhyā, Series A*, **59**, 311–323.

Dallas, A. C. (1973). A characterization of the exponential distribution. *Bulletin of the Greek Society of Mathematics*, **14**, 172–175.

Dembinska, A. and Wesolowski, J. (2000). Linearity of regression for non-adjacent record values. *Journal of Statistical Planning and Inference*, **90**, 195–205.

Ferguson, T. S. (1967). On characterizing distributions by properties of order statistics. *Sankhyā, Series A*, **29**, 265–278.

Nagaraja, H. N. (1988a). Some characterizations of discrete distributions based on linear regressions of adjacent order statistics. *Journal of Statistical Planning and Inference*, **20**, 65–75.

Nagaraja, H. N. (1988b). Some characterizations of continuous distributions based on regression of adjacent order statistics and record values. *Sankhyā, Series A*, **50**, 70–73.

Wang, Y. H. and Srivastava, R. C. (1980). A characterization of the exponential and related distributions by linear regression. *Annals of Statistics*, **8**, 217–220.

Wesolowski, J. and Ahsanullah, M. (1997). On characterizing distributions via linearity of regression for order statistics. *Australian Journal of Statistics*, **39**, 69–78.

Chapter 5

Characterization of the Exponential Distribution by Conditional Expectations of Generalized Spacings

ERHARD CRAMER AND UDO KAMPS
Institute of Statistics,
RWTH Aachen University
Aachen, Germany

CONTENTS

ABSTRACT

A characterization of the exponential distribution based on a conditional expectation

$$E\left(X_*^{(r+l+v)} - X_*^{(r+v)} | X_*^{(v)}\right) = b \quad P^F \text{ a.e.}$$

is presented where $X_*^{(1)}, \ldots, X_*^{(n)}$ are generalized order statistics based on an absolutely continuous distribution function F.

KEYWORDS AND PHRASES: Characterization, exponential distribution, generalized order statistics, order statistics, spacings

5.1   INTRODUCTION

Characterizations of the exponential distributions based on order statistics and record values have been considered extensively in the literature. For surveys we refer to Azlarov and Volodin (1986), Arnold and Balakrishnan (1989), Arnold and Huang (1995), and Chapters 8 to 10 in Balakrishnan and Rao (1998). Here, we present a characterization based on conditional expectations of generalized order statistics. Namely, a constant conditional expectation

$$E\left(X_*^{(r+l+\nu)} - X_*^{(r+\nu)}|X_*^{(\nu)}\right) = b \quad P^F \text{ a.e.}$$

for generalized order statistics $X_*^{(1)}, \ldots, X_*^{(n)}$ based on an absolutely continuous F implies that F is exponential.

Uniform generalized order statistics $U_*^{(1)}, \ldots, U_*^{(n)}$ based on parameters $\gamma_1, \ldots, \gamma_n > 0$ are introduced via the joint density function

$$f^{U_*^{(1)},\ldots,U_*^{(n)}}(u_1, \ldots, u_n)$$

$$= \left(\prod_{j=1}^{n} \gamma_j\right) \left(\prod_{j=1}^{n-1}(1 - u_j)^{m_j}\right) (1 - u_n)^{\gamma_n-1},$$

$$0 < u_1 \leq \cdots \leq u_n < 1$$

with $n \in \mathbb{N}$, $n \geq 2$, and $m_j = \gamma_j - \gamma_{j+1} - 1$, $1 \leq j \leq n-1$ [cf. Kamps (1995a,b)]. Using the quantile function F^{-1} of a distribution function F, generalized order statistics $X_*^{(1)}, \ldots, X_*^{(n)}$ based on F and parameters $\gamma_1, \ldots, \gamma_n > 0$ are defined via $X_*^{(j)} = F^{-1}(U_*^{(j)})$, $1 \leq j \leq n$. In the following, particular generalized order statistics with $m_j = m \in \mathbb{R}$, $1 \leq j \leq n-1$, i.e., $\gamma_j = k + (n-j)(m+1) > 0$, $1 \leq j \leq n$, $(\gamma_n = k)$ are called m-generalized order statistics.

Choosing particular values for the parameters, well-known models result, e.g., order statistics, record values, and progressively type II censored order statistics [cf. Cramer and Kamps (2001)]. It has been shown in Cramer (2002) and Cramer and Kamps (2003) that the set-up of generalized order statistics leads to a unified distribution theory for all these models. Characterizations of distributions based on generalized order statistics are presented in, e.g., Ahsanullah (1995), Kamps (1995a, 1996), Kamps and Gather (1997), Ahsanullah and Nevzorov (2001), Kamps and Keseling (2002), and Cramer et al. (2003a,b). In particular, Cramer et al. (2003a) establish a general result that characterizes generalized Pareto distributions by the linearity of regression. In particular, this type of characterization has been considered for order statistics by, e.g., Ferguson (1967), Pudeg (1991), Franco and Ruiz (1995, 1996), López Blázquez and Moreno Rebollo (1997), and Dembińska and Wesołowski (1998); for record values by Nagaraja (1977, 1988), Grudzień and Szynal (1985), and Dembińska and Wesołowski (2000), and for *m*-generalized order statistics by Keseling (1999a,b). Many of these results can be seen as a particular case of the following theorem [cf. Cramer et al. (2003a)].

THEOREM 5.1.1 *Let* $1 \le r \le n - 1$, $1 \le l \le n - r$, *and* $X_*^{(1)}, \ldots,$ $X_*^{(n)}$ *be generalized order statistics based on a continuous distribution function* F *and parameters* $\gamma_1, \ldots, \gamma_n$.
If constants $a > 0$ *and* $b \in \mathbb{R}$ *exist such that*

$$E\left(X_*^{(r+l)} \mid X_*^{(r)} = x\right) = ax + b \quad P^F \text{ a.e.}$$

then F *is the distribution function of a generalized Pareto distribution. Thus, up to an affine transformation of the argument, it is given by one of the following distribution functions.*

1. *If* $a = 1$, *then*

$$F(x) = 1 - \exp(-x), \quad x \ge 0.$$

2. *If* $0 < a < 1$, *then*

$$F(x) = 1 - (-x)^\theta, \quad x \in [-1, 0].$$

3. *If* $1 < a$, *then*

$$F(x) = 1 - x^\theta, \quad x \in [1, \infty).$$

The parameter θ is given by $\theta = -\frac{1}{\eta}$ where η is the unique solution of the polynomial equation

$$a \prod_{j=r+1}^{r+l} (\gamma_j - \eta) = \prod_{j=r+1}^{r+l} \gamma_j, \quad \eta \in (-\infty, \min\{\gamma_{r+1}, \dots, \gamma_{r+l}\}).$$

For $a = 1$, Theorem 5.1.1 yields a characterization of the exponential distribution which can be reformulated as

$$E\left(X_*^{(r+l)} - X_*^{(r)} | X_*^{(r)} = \cdot\right) = b \quad P^F \text{ a.e.} \tag{5.1.1}$$

Hence, the conditional expectation of the spacing $X_*^{(r+l)} - X_*^{(r)}$ given $X_*^{(r)}$ is constant. In this paper, we extend this result in Theorem 5.3.1 to the conditional expectation of $X_*^{(r+l)} - X_*^{(r)}$ given $X_*^{(s)}$ with $s < r$.

5.2   PRELIMINARIES

The proof of the main result in Theorem 5.3.1 is based on the distribution theory for generalized order statistics presented in Cramer (2002) and Cramer and Kamps (2003). In particular, generalized order statistics $X_*^{(1)}, \dots, X_*^{(n)}$ are distributed as

$$F^{-1}(1 - B_1), \dots, F^{-1}\left(1 - \prod_{j=1}^{n} B_j\right),$$

where $B_1, \dots, B_n$ are independent, power-function-distributed random variables with distribution function $F^{B_j}(t) = t^{\gamma_j}$, $t \in [0, 1]$, $1 \leq j \leq n$. This expression yields, for instance, that the distribution function of a generalized order statistic $X_*^{(r)}$ has the integral representation

$$F_{*,r}(t) = 1 - \left(\prod_{j=1}^{r} \gamma_j\right) \int_0^{1-F(t)} \mathbf{G}_{r,r}^{r,0} [x \,|\gamma_1, \dots, \gamma_r] \, dx,$$

where

$$\mathbf{G}_{r,r}^{r,0}\,[x\,|\gamma_1,\ldots,\gamma_r\,] = \mathbf{G}_{r,r}^{r,0}\left[x\left|\begin{array}{ccc}\gamma_1, & \ldots, & \gamma_r \\ \gamma_1-1, & \ldots, & \gamma_r-1\end{array}\right.\right],\, x\in(0,1),$$

denotes a particular Meijer's G-function [for details on Meijer's G-functions see Mathai (1993)]. Given that F is continuous, it is shown in Lemma 3.1.9 of Cramer (2002) that P^F and $P^{X_*^{(r)}}$ are equivalent measures such that $P^{X_*^{(r)}}$ has a Radon-Nikodym derivative w.r.t. P^F, i.e.,

$$\frac{dP^{X_*^{(r)}}}{dP^F}(t) = \left(\prod_{j=1}^{r}\gamma_j\right)\mathbf{G}_{r,r}^{r,0}\,[1-F(t)\,|\gamma_1,\ldots,\gamma_r\,]$$

$$\text{a.e.,}\ t\in\mathbb{R}\quad(5.2.1)$$

[cf. Keseling (1999a) for m-generalized order statistics]. In particular, a conditional P^F-density function of $X_*^{(r+l)}$ given $X_*^{(r)} = x,\, l\geq 1$, can be written as

$$f^{X_*^{(r+l)}|X_*^{(r)}}(t|x) = \left(\prod_{v=r+1}^{r+l}\gamma_v\right)\frac{1}{1-F(x)}$$

$$\times\,\mathbf{G}_{l,l}^{l,0}\,[1-F_x(t)\,|\gamma_{r+1},\ldots,\gamma_{r+l}\,]\,\mathbf{1}_{(x,\omega(F))}(t),$$

$$x,t\in\mathbb{R},$$

where $\omega(F) = F^{-1}(1)$ and

$$F_x(t) = \begin{cases} \frac{F(t)-F(x)}{1-F(x)}, & t\geq x \\ 0, & t< x \end{cases},\qquad x<\omega(F).\quad(5.2.2)$$

These expressions show that the measure $P^{X_*^{(r+l)}|X_*^{(r)}=x}$ can be seen as the distribution of a generalized order statistic based on the left truncated distribution function F_x and parameters $\gamma_{r+1},\ldots,\gamma_{r+l}$ [see Cramer et al. (2003a) for generalized order statistics and Arnold et al. (1992, Theorem 2.4.1, p. 23) for ordinary order statistics].

Meijer's G-functions satisfy many important relations [see Cramer (2002), Cramer and Kamps (2003), and Mathai (1993)]. In the proof of Theorem 5.3.1, we make use of the following two identities.

LEMMA 5.2.1 *Let $r \geq 2$ and $z \in (0, 1]$.*
    Then,

(i) $\dfrac{d}{dz}\, \mathbf{G}_{r,r}^{r,0}\, [z\,|\gamma_1, \ldots, \gamma_r\,]$

$$= \frac{1}{z}\Big((\gamma_r - 1)\, \mathbf{G}_{r,r}^{r,0}\, [z\,|\gamma_1, \ldots, \gamma_r\,]$$

$$- \mathbf{G}_{r-1,r-1}^{r-1,0}\, [z\,|\gamma_1, \ldots, \gamma_{r-1}]\,\Big),$$

(ii) $z^a\, \mathbf{G}_{r,r}^{r,0}\, [z\,|\gamma_1, \ldots, \gamma_r\,] = \mathbf{G}_{r,r}^{r,0}\, [z\,|\gamma_1 + a, \ldots, \gamma_r + a],$
$$a \in \mathbb{R}.$$

5.3 CHARACTERIZATION RESULT

As mentioned above, the characterization of the exponential distribution given in Theorem 5.1.1 can be rewritten as in (5.1.1) with some constant $b > 0$. This observation leads us to a characterization of the exponential distribution in terms of more general regressions. Namely, the exponential distribution is characterized by the equation

$$E\left(X_*^{(r+l+\nu)} - X_*^{(r+\nu)}|X_*^{(\nu)} = \cdot\right) = b \quad P^F \text{ a.e.}$$

for some r, l, ν. The following characterization result extends Theorem 3.8 of Keseling (1999a) which is restricted to the case of different parameters of the generalized order statistics. A similar result for record values was given by Huang and Li (1993).

THEOREM 5.3.1 *Let $X_*^{(1)}, \ldots, X_*^{(n)}$ be generalized order statistics based on an absolutely continuous distribution function F with density function f and parameters $\gamma_1, \ldots, \gamma_n$. Suppose that for numbers $r, l, \nu \in \mathbb{N}$ with $r + l + \nu \leq n$ the expectation $E\,|X_*^{(r+l+\nu)}|$ is finite.*

If a constant b > 0 exists such that

$$E\left(X_*^{(r+l+v)} - X_*^{(r+v)} | X_*^{(v)} = \cdot\right) = b \quad P^F \text{ a.e.}$$

then

(i)   *for any $j \in \{1, \ldots, r\}$, the following relation holds:*

$$E\left(X_*^{(r+l+v)} - X_*^{(r+v)} | X_*^{(v+j)} = \cdot\right) = b \quad P^F \text{ a.e.}$$

(ii)   *F is the distribution function of a two-parameter exponential distribution.*

PROOF. In the case $j = r$ in (i), we can apply Theorem 5.1.1 yielding (ii).

In order to prove (i), we establish the result first for $j = 1$.

Let $\alpha(F) = F^{-1}(0+)$, $\omega(F) = F^{-1}(1)$. For $x \in (\alpha(F), \omega(F))$, let $Z_x^{(r+l)}, Z_x^{(r)}$ be generalized order statistics based on F_x and parameters $\gamma_{v+1}, \ldots, \gamma_{r+l+v}$ and $\gamma_{v+1}, \ldots, \gamma_{r+v}$, respectively [cf. (5.2.2)]. The density function of F_x is denoted by $f_x = f/(1 - F(x))$. Using the random variables $Z_x^{(r+l)}, Z_x^{(r)}$, the conditional expectation can be written as [cf. Cramer et al. (2003a)]

$$b = E\left(X_*^{(r+l+v)} - X_*^{(r+v)} | X_*^{(v)} = x\right)$$

$$= E(Z_x^{(r+l)} - Z_x^{(r)}) \quad P^F \text{ a.e.} \tag{5.3.1}$$

Since the joint $P^F \otimes P^F$-density of $X_*^{(r)}$ and $X_*^{(r+l)}$ is given by

$$f^{X_*^{(r)}, X_*^{(r+l)}}(x, t) = \left(\prod_{v=1}^{r+l} \gamma_v\right) \frac{1}{1 - F(x)}$$

$$\times G_{l,l}^{l,0}\left[\frac{1 - F(t)}{1 - F(x)} \Big| \gamma_{r+1}, \ldots, \gamma_{r+l}\right]$$

$$\times G_{r,r}^{r,0}\left[1 - F(x) | \gamma_1, \ldots, \gamma_r\right], \tag{5.3.2}$$

where $\alpha(F) < x \leq t < \omega(F)$, the joint density (w.r.t. λ^2) of $Z_x^{(r)}$ and $Z_x^{(r+l)}$ is given by

$$f^{Z_x^{(r)}, Z_x^{(r+l)}}(z, t)$$

$$= \left(\prod_{j=v+1}^{r+l+v} \gamma_j\right) f_x(z) f_x(t) \frac{1}{1 - F_x(z)}$$

$$\times \mathbf{G}_{l,l}^{l,0}\left[\frac{1 - F_x(t)}{1 - F_x(z)}\Big|\gamma_{r+v+1}, \ldots, \gamma_{r+v+l}\right]$$

$$\times \mathbf{G}_{r,r}^{r,0}\left[1 - F_x(z)\Big|\gamma_{1+v}, \ldots, \gamma_{r+v}\right]$$

$$= \left(\prod_{j=v+1}^{r+l+v} \gamma_j\right) f_x(z) f_x(t) \frac{1}{1 - F_x(z)}$$

$$\times \mathbf{G}_{l,l}^{l,0}\left[1 - F_z(t)\Big|\gamma_{r+v+1}, \ldots, \gamma_{r+v+l}\right]$$

$$\times \mathbf{G}_{r,r}^{r,0}\left[1 - F_x(z)\Big|\gamma_{1+v}, \ldots, \gamma_{r+v}\right].$$

Hence, a λ^1-density of $W_x = Z_x^{(r+l)} - Z_x^{(r)}$ is specified by

$$f^{W_x}(t) = \left(\prod_{j=v+1}^{r+l+v} \gamma_j\right) \int_{(x,\infty)} \frac{1}{1 - F_x(z)}$$

$$\times \mathbf{G}_{l,l}^{l,0}\left[1 - F_z(z+t)\Big|\gamma_{r+v+1}, \ldots, \gamma_{r+v+l}\right]$$

$$\times \mathbf{G}_{r,r}^{r,0}\left[1 - F_x(z)\Big|\gamma_{1+v}, \ldots, \gamma_{r+v}\right]$$

$$\times f_x(z+t) f_x(z) \, d\lambda^1(z), \quad t \geq 0.$$

Considering the expectation $EW_x = \int t f^{W_x}(t) \, d\lambda^1(t)$, and the definition of f_x, and applying Fubini's theorem, we obtain the equation

$$\left(\prod_{j=v+1}^{r+l+v} \gamma_j\right) \frac{1}{1 - F(x)}$$

$$\times \int_{(x,\infty)} \mathbf{G}_{r,r}^{r,0}\left[1 - F_x(z)\Big|\gamma_{1+v}, \ldots, \gamma_{r+v}\right]$$

$$\times f(z) g(z) \, d\lambda^1(z) = b \quad P^F \quad \text{a.e.,} \tag{5.3.3}$$

where

$$g(z) = \int_{(0,\infty)} t \, \mathbf{G}_{l,l}^{l,0} \left[1 - F_z(z+t) \,\middle|\, \gamma_{r+v+1}, \ldots, \gamma_{r+v+l}\right]$$
$$\times f_z(z+t) \, d\lambda^1(t).$$

Let N be the set of values such that (5.3.3) does not hold. Thus, $P^F(N) = 0$. Since both sides of the equation are continuous in x, the equality can be extended to any limit point of N. If $N \cap (\alpha(F), \omega(F))$ has an inner point x_0, then $\varepsilon > 0$ and a constant $c \in \mathbb{R}$ exist with

$$EW_x = c \quad \text{for all } x \in (x_0 - \varepsilon, x_0 + \varepsilon) \subseteq N.$$

Choosing ε maximal in the sense that $x_1 = x_0 - \varepsilon \in N^c$ or $x_2 = x_0 + \varepsilon \in N^c$ and $(x_1, x_2) \subseteq N$, we find $EW_{x_1} = b$ or $EW_{x_2} = b$, respectively. Since the integral on the left-hand side of (5.3.3) is continuous in x, this yields $c = b$. Hence, the equation holds for any $x \in (\alpha(F), \omega(F))$.

Now, we multiply both sides of (5.3.3) by $(1 - F(x))^{\gamma_{v+1}}$ and differentiate them w.r.t. x. The result on the right-hand side is given by $b\gamma_{v+1} f(x)(1 - F(x))^{\gamma_{v+1}-1} \lambda^1$ a.e. The derivative of the left-hand side equals

$$\left(\prod_{j=v+1}^{r+l+v} \gamma_j \right) \Bigg[(1 - F(x))^{\gamma_{v+1}-1}$$
$$\times \mathbf{G}_{r,r}^{r,0} \left[1 - F_x(x) \,\middle|\, \gamma_{1+v}, \ldots, \gamma_{r+v}\right] f(x)g(x)$$
$$- \int_{(x,\infty)} \frac{d}{dx} \Big\{ (1 - F(x))^{\gamma_{v+1}-1}$$
$$\times \mathbf{G}_{r,r}^{r,0} \left[1 - F_x(z) \,\middle|\, \gamma_{1+v}, \ldots, \gamma_{r+v}\right] \Big\} f(z)g(z) d\lambda^1(z) \Bigg].$$

Since $1 - F_x(x) = 1$ and $\mathbf{G}_{r,r}^{r,0} \left[1 \,\middle|\, \gamma_{v+1}, \ldots, \gamma_{r+v}\right] = 0$, the first term cancels out. The derivative of the term $\{\ldots\}$ is calculated as follows. The factor $(1 - F(x))^{\gamma_{v+1}-1}$ can be written as

$$(1 - F(x))^{\gamma_{v+1}-1} = (1 - F_x(z))^{1-\gamma_{v+1}}(1 - F(z))^{1-\gamma_{v+1}}.$$

$$(5.3.4)$$

Using Lemma 5.2.1 (ii), this leads to

$$(1 - F_x(z))^{1-\gamma_{v+1}} \, \mathbf{G}_{r,r}^{r,0} \left[1 - F_x(z) \, \middle| \, \gamma_{1+v}, \ldots, \gamma_{r+v} \right]$$

$$= \mathbf{G}_{r,r}^{r,0} \left[1 - F_x(z) \, \middle| \, 1, \gamma_{2+v} - \gamma_{1+v} + 1, \ldots, \gamma_{r+v} - \gamma_{1+v} + 1 \right].$$

Differentiating this expression w.r.t x and applying Lemma 5.2.1 (ii) in the reverse direction, this yields by Lemma 5.2.1 (i) the expression

$$\frac{d}{dx} \, \mathbf{G}_{r,r}^{r,0} \left[1 - F_x(z) \, \middle| \, 1, \gamma_{2+v} - \gamma_{1+v} + 1, \ldots, \gamma_{r+v} - \gamma_{1+v} + 1 \right]$$

$$= \frac{1}{1 - F_x(z)} \frac{f(x)(1 - F(z))}{(1 - F(x))^2}$$

$$\times \, \mathbf{G}_{r-1,r-1}^{r-1,0} \left[1 - F_x(z) \, \middle| \, \gamma_{2+v} - \gamma_{1+v} + 1, \right.$$

$$\left. \ldots, \gamma_{r+v} - \gamma_{1+v} + 1 \right]$$

$$= \frac{f(x)}{1 - F(x)} (1 - F_x(z))^{1-\gamma_{v+1}}$$

$$\times \, \mathbf{G}_{r-1,r-1}^{r-1,0} \left[1 - F_x(z) \, \middle| \, \gamma_{2+v}, \ldots, \gamma_{r+v} \right].$$

Hence, we obtain the equation

$$b\gamma_{v+1} f(x)(1 - F(x))^{\gamma_{v+1}-1}$$

$$= \left(\prod_{j=v+1}^{r+l+v} \gamma_j \right) \int_{(x,\infty)} \frac{f(x)}{1 - F(x)} (1 - F_x(z))^{1-\gamma_{v+1}}$$

$$\times \, \mathbf{G}_{r-1,r-1}^{r-1,0} \left[1 - F_x(z) \, \middle| \, \gamma_{2+v}, \ldots, \gamma_{r+v} \right]$$

$$\times (1 - F(z))^{1-\gamma_{v+1}} f(z) g(z) d\lambda^1(z)$$

valid λ^1 a.e. on $(\alpha(F), \omega(F))$. Simplifying the right-hand side according to (5.3.4) leads to

$$\left(\prod_{j=v+1}^{r+l+v} \gamma_j \right) f(x)(1 - F(x))^{\gamma_{v+1}-2}$$

$$\times \int_{(x,\infty)} \mathbf{G}_{r-1,r-1}^{r-1,0} \left[1 - F_x(z) \,|\, \gamma_{2+v}, \ldots, \gamma_{r+v} \right]$$

$$\times f(z)g(z)\,d\lambda^1(z).$$

Dividing both sides of the resulting equation by $\gamma_{v+1}f(x)(1 - F(x))^{\gamma_{v+1}-1}$ yields the equation

$$b = \left(\prod_{j=v+2}^{r+l+v} \gamma_j \right) \frac{1}{1 - F(x)}$$

$$\times \int_{(x,\infty)} \mathbf{G}_{r-1,r-1}^{r-1,0} \left[1 - F_x(z) \,|\, \gamma_{2+v}, \ldots, \gamma_{r+v} \right]$$

$$\times f(z)g(z)\,d\lambda^1(z) \quad P^F \text{ a.e.}$$

However, this is equation (5.3.3) with $v + 1$ and $r - 1$ instead of v and r, respectively. The parameters of the respective generalized order statistics are $\gamma_{v+2}, \ldots, \gamma_{r+l+v}$. Repeating the calculations to obtain (5.3.3) backwards, we arrive at (5.3.1):

$$b = E\left(X_*^{(r+l+v)} - X_*^{(r+v)} | X_*^{(v+1)} = x \right) \quad P^F \text{ a.e.,}$$

where the generalized order statistics are based on F and parameters $\gamma_{v+2}, \ldots, \gamma_{r+l+v}$. Note that the conditional expectation is independent of the first $v+1$ parameters. This proves (i), for $j = 1$. Using an induction argument, we obtain the result for any $j \in \{1, \ldots, r\}$ and, hence, the assertion. ∎

REFERENCES

Ahsanullah, M. (1995). The generalized order statistics and a characteristic property of the exponential distribution. *Pakistan Journal of Statistics*, **11**, 215–218.

Ahsanullah, M. and Nevzorov, V. (2001). Distributions between uniform and exponential distribution. In *Applied Statistical Science V* (Eds., M. Ahsanullah, J. Kenyon, and S. K. Sarkar). Nova Science Publishers, Commack, New York.

Arnold, B. C. and Balakrishnan, N. (1989). *Relations, Bounds and Approximations for Order Statistics*. Lecture Notes in Statistics Vol. 53. Springer-Verlag, New York.

Arnold, B. C., Balakrishnan, N. and Nagaraja, H. N. (1992). *A First Course in Order Statistics*. John Wiley & Sons, New York.

Arnold, B. C. and Huang, J. S. (1995). Characterizations. In *The Exponential Distribution: Theory, Methods and Applications* (Eds., N. Balakrishnan and A. P. Basu), pp. 185–203. Gordon and Breach Science Publishers, Newark, New Jersey.

Azlarov, T. A. and Volodin, N. A. (1986). *Characterization Problems Associated with the Exponential Distribution*. Springer-Verlag, New York.

Balakrishnan, N. and Rao, C. R. (Eds.) (1998). *Order Statistics: Theory and Methods*. Handbook of Statistics, Vol. 16. North-Holland, Amsterdam.

Cramer, E. (2002). *Contributions to Generalized Order Statistics*. Habilitationsschrift, University of Oldenburg, Oldenburg, Germany.

Cramer, E. and Kamps, U. (2001). Sequential k-out-of-n systems. In *Handbook of Statistics, Vol. 20: Advances in Reliability* (Eds., N. Balakrishnan and C. R. Rao), pp. 301–372. North-Holland, Amsterdam.

Cramer, E. and Kamps, U. (2003). Marginal distributions of sequential and generalized order statistics. *Metrika* **58**, 293–310.

Cramer, E., Kamps, U. and Keseling, C. (2003a). Characterizations via linear regression of ordered random variables: a unifying approach. *Communications in Statistics-Theory and Methods*, **33**, 2885–2912.

Cramer, E., Kamps, U. and Raqab, M. Z. (2003b). Characterizations of exponential distributions by spacings of generalized order statistics. *Applications Mathematicae* **36**, 257–265.

Dembińska, A. and Wesołowski, J. (1998). Linearity of regression from non-adjacent order statistics. *Metrika*, **90**, 215–222.

Dembińska, A. and Wesołowski, J. (2000). Linearity of regression from non-adjacent record values. *Journal of Statistical Planning and Inference*, **90**, 195–205.

Ferguson, T. S. (1967). On characterizing distributions by properties of order statistics. *Sankhyā, Series A*, **29**, 265–278.

Franco, M. and Ruiz, J. M. (1995). On characterization of continuous distributions with adjacent order statistics. *Statistics*, **26**, 375–385.

Franco, M. and Ruiz, J. M. (1996). On characterization of continuous distributions by conditional expectation of record values. *Sankhyā, Series A*, **58**, 135–141.

Grudzień, Z. and Szynal, D. (1985). On the expected values of *k*th record values and associated characterizations of distributions. In *Probability and Statistical Decision Theory, Vol. A* (Eds., F. Konecny, J. Mogyoródi, and W. Wertz), pp. 119–127, Reidel, Dordrecht, The Netherlands.

Huang, W. J. and Li, S. H. (1993). Characterization results based on record values. *Statistica Sinica*, **3**, 583–599.

Kamps, U. (1995a). *A Concept of Generalized Order Statistics*. Teubner, Stuttgart, Germany.

Kamps, U. (1995b). A concept of generalized order statistics. *Journal of Statistical Planning and Inference*, **48**, 1–23.

Kamps, U. (1996). A characterization of uniform distributions by sub-ranges and its extension to generalized order statistics. *Metron*, **54**, 37–44.

Kamps, U. and Gather, U. (1997). Characteristic properties of generalized order statistics from exponential distributions. *Applied Mathematics*, **24**, 383–391.

Kamps, U. and Keseling, C. (2003). A theorem of Rossberg for generalized order statistics. *Sankhyā, Series A* **65**, 259–270.

Keseling, C. (1999a). *Characterizations of Probability Distributions by Generalized Order Statistics (in German)*. Ph.D. thesis, Aachen University of Technology, Aachen, Germany.

Keseling, C. (1999b). Conditional distributions of generalized order statistics and some characterizations. *Metrika*, **49**, 27–40.

López Blázquez, F. and Moreno Rebollo, J. L. (1997). A characterization of distributions based on linear regression of order statistics and record values. *Sankhyā, Series A*, **59**, 311–323.

Mathai, A. M. (1993). *A Handbook of Generalized Special Functions for Statistical and Physical Sciences*. Clarendon Press, Oxford, U.K.

Nagaraja, H. N. (1977). On a characterization based on record values. *Australian Journal of Statistics*, **19**, 70–73.

Nagaraja, H. N. (1988). Some characterizations of continuous distributions based on regressions of adjacent order statistics and record values. *Sankhyā, Series A*, **50**, 70–73.

Pudeg, A. (1991). *Characterizations of Probability Distributions by Distributional Properties of Order Statistics and Records (in German)*. Ph.D. thesis, Aachen University of Technology, Aachen, Germany.

Chapter 6

Some Characterizations of Exponential Distribution Based on Progressively Censored Order Statistics

N. BALAKRISHNAN
Department of Mathematics and Statistics,
McMaster University, Hamilton, Ontario, Canada

S. V. MALOV[1]
Department of Mathematics and Mechanics,
St. Petersburg State University, St. Petersburg, Russia

CONTENTS

[1] Partially supported by the Russian Foundation of Basic Research (grant N HIII-2258.2003.1) and the Natural Sciences and Engineering Research Council of Canada.

ABSTRACT

We consider some characterizations of the exponential distribution via the independence property of spacings of Type-II progressively censored order statistics. We also give characterizations of the class via regression properties of these order statistics.

KEYWORDS AND PHRASES: Characterizations, exponential distribution, order statistics, Type-II progressively censored order statistics

6.1   INTRODUCTION

Let $X_1, \ldots, X_n$ be independent and identically distributed random variables having a distribution function F; let $X_{(1)} \leq \ldots \leq X_{(n)}$ be the corresponding order statistics and $\Delta_1, \ldots, \Delta_n$ be the spacings. Suppose also that $R_1, \ldots, R_m$ are some fixed non-negative integers such that $\sum_{i=1}^{m} R_i = n - m$. Consider the Type-II progressively censored order statistics $X_{(1)}^* \leq \ldots \leq X_{(m)}^*$ generated in the following manner. Let $X_{(1)}^* = X_{(1)}$. After that, we remove randomly R_1 elements from $X_{(2)}, \ldots, X_{(n)}$. Denote the order statistics among those left by $X_{(1)}^{(2)}, \ldots, X_{(n-\alpha_1)}^{(2)}$, and let $X_{(2)}^* = X_{(1)}^{(2)}$. Now we remove R_2 elements from $X_{(2)}^{(2)}, \ldots,$ $X_{(n-\alpha_1)}^{(2)}$ randomly and continue the process by taking $X_{(i)}^* = X_{(1)}^{(i)}$ for $i = 2, 3, \ldots, m$ and removing R_i elements from those left at $X_{(i)}^*$, where $\alpha_i = i + R_1 + \cdots + R_i$ (with $\alpha_0 = 0$). As was observed by Balakrishnan and Aggarwala (2000) and Kamps (1995), the distribution of $X_{(1)}^*, \ldots, X_{(m)}^*$ has some remarkable properties which are similar to the properties of the usual order statistics. It is easy to see that $X_{(i)}^* = X_{(i)}$ for all $i = 1, \ldots, k+1$ under the condition $R_1 = \ldots = R_k = 0$, $k \leq m < n$, and under the condition $k = m = n$ we obtain the usual order statistics. On the other hand, in the case when $R_1 = \ldots = R_k = R$, $k \leq l$ and $Rl = n$, there exists the following representation of the distribution of the vector $X_{(1)}^*, \ldots, X_{(k)}^*$. Let $Y_i = \min \{X_j, j = (i-1)(R+1)+1, \ldots, i(R+1)\}$, $i = 1, \ldots, l$, and $Y_{(1)}, \ldots, Y_{(l)}$ be the corresponding order statistics. Then by the exchangeability of the initial random variables, we can conclude that

$$(Y_{(1)}, \ldots, Y_{(k)}) \overset{\mathrm{d}}{=} (X^*_{(1)}, \ldots, X^*_{(k)}).$$

Introduce also the corresponding spacings $\Delta^*_1 = X^*_{(1)}$ and $\Delta^*_i = X^*_{(i)} - X^*_{(i-1)}$, $i = 2, \ldots, m$. We consider here generalizations of some characterizations of exponential distribution via independence and linear regression properties of spacings and order statistics. The characterization of exponential distribution when $n = 2$ by independence property of spacing Δ_2 and the minimum $X_{(1)}$ was given by Fisz (1958). Rogers (1959) proved that if $X_1, \ldots, X_n$ are i.i.d. random variables, then the independence of the order statistic $X_{(h)}$ on some random variable $\phi(X_{(k)}, \ldots, X_{(n)})$, $h \leq k$, implies the independence $X_{(k)}$ with this random variable. Rossberg (1972) proved that, under some additional restrictions on the distribution function F, an analogous property holds if one changes $X_{(h)}$ by $\Delta_{(h)}$. Rossberg (1972) also gave the characterization of the exponential distribution via independence property of $X_{(k)}$ and some linear combination of spacings $c_{k+1} \Delta_{k+1} + \cdots + c_n \Delta_n = s_k X_{(k)} + \cdots + s_n X_{(n)}$, where $s_k + \cdots + s_n = 0$. Some characterizations of the exponential distribution via conditional expectation of spacing (conditioned on the nearest order statistic) were given by Rogers (1963), Beg and Kirmani (1989), and Rao and Shanbhag (1994).

6.2 THE CHARACTERIZATIONS

To begin with, we formulate some results which we will use to modify the above-mentioned characterization theorems. The central part of this process is the Markovian property of the sequence $X^*_{(1)}, \ldots, X^*_{(m)}$ and the following representations [see, for example, Balakrishnan and Aggarwala (2000)].

REPRESENTATION 6.2.1 *Let* $(X^*_{(1)}, \ldots, X^*_{(m)})$ *be Type-II progressively censored order statistics, and* $k < m$. *Then the conditional distribution of Type-II progressively censored order statistics* $X^*_{(k+1)}, \ldots, X^*_{(m)}$, *given* $X^*_{(1)}, \ldots, X^*_{(k)}$, *coincides with the distribution of Type-II progressively censored order statistics arising* *from a sample of size $\beta_k = n - \alpha_k$ with $R_i(k) = R_{i+k}$, $i = 1, \ldots,$ $m - k$, from the left-truncated distribution with cdf $G_{(k)}(x) = \frac{F(x)}{1 - F(X^*_{(k)})} \mathbb{I}_{\{x \geq X^*_{(k)}\}}.$*

REPRESENTATION 6.2.2 *Let $(X^*_{(1)}, \ldots, X^*_{(m)})$ be Type-II progressively censored order statistics from a sample from the standard exponential distribution. Then the corresponding spacings $\Delta^*_1, \ldots, \Delta^*_m$ are independent random variables having distribution functions $F_{\Delta^*_i}(x) = \{1 - \exp(-\beta_{i-1}x)\} \mathbb{I}_{\{x > 0\}}.$*

To improve the results of Rogers (1959) and Rossberg (1972), we use the following simple lemma, which is, in fact, a version of the well-known theorem of Basu (1955).

LEMMA 6.2.1 *Let $\xi_1, \ldots, \xi_n$ be random variables satisfying the Markovian property. Suppose for $1 \leq h \leq k \leq l \leq n$ and for any real-valued function $g \in \Re$, the condition*

$$\mathbf{E}[g(\xi_k)|\phi(\xi_1, \ldots, \xi_h) = s] = 0 \quad \text{for any } s \text{ a.s. implies that}$$

$$g(\xi_k) = 0 \text{ a.s.} \tag{6.2.1}$$

Then for any function f such that $g(\xi_k) = \mathbf{E}[f(\xi_l, \ldots, \xi_n)|\xi_k] - \mathbf{E} f(\xi_l, \ldots, \xi_n)$ is in $\Re$ a.s. and $\mathbf{E}[f(\xi_l, \ldots, \xi_n)|\phi(\xi_1, \ldots, \xi_h)]$ is non-random, the random variable $\mathbf{E}[f(\xi_l, \ldots, \xi_n)|\xi_k]$ must also be non-random.

PROOF. By the Markovian property, under the assumptions of the lemma, we have

$$\mathbf{E}[f(\xi_l, \ldots, \xi_n)|\phi(\xi_1, \ldots, \xi_h)] - \mathbf{E}[f(\xi_l, \ldots, \xi_n)]$$

$$= \int \mathbf{E}[f(\xi_l, \ldots, \xi_n)|\xi_k = x, \phi(\xi_1, \ldots, \xi_h)]$$

$$\times F_{\xi_k|\phi}(dx) - \mathbf{E} f(\xi_l, \ldots, \xi_n)$$

$$= \int (\mathbf{E}[f(\xi_l, \ldots, \xi_n)|\xi_k = x] - \mathbf{E}[f(\xi_l, \ldots, \xi_n)])$$

$$\times F_{\xi_k|\phi}(dx) = 0.$$

Therefore, by Condition (6.2.1), we obtain

$$\mathbf{E}[f(\xi_l, \ldots, \xi_n)|\xi_k] = \mathbf{E}[f(\xi_l, \ldots, \xi_n)] \quad \text{a.s.}$$

Hence, the lemma is proved. ∎

The following lemma is an improvement of the result of Rogers (1959).

LEMMA 6.2.2 *Let* $X_1, \ldots, X_n$ *be i.i.d. random variables and* $R_1, \ldots, R_m \geq 0$ *be some non-negative integers such that* $n - m = \sum_{i=1}^m R_i$. *Then for any* $1 \leq h \leq k \leq m$, *the condition* $\mathbf{E}[f(Z)|X_{(h)}^*] = c = constant$, *where* $Z = \phi(X_{(k)}^*, \ldots, X_{(m)}^*)$, *implies that* $\mathbf{E}[f(Z)|X_{(k)}^*] = \mathbf{E}[f(Z)|X_{(1)}^*, \ldots, X_{(k)}^*] = c$.

PROOF. Rogers (1959) has proved that Condition (6.2.1) holds for the usual order statistics, where $\Re$ is a class of bounded measurable functions. We use similar arguments for the case of Type-II progressively censored order statistics and all measurable $d\bar{F}^{\beta_k}$-integrable functions g, where $\bar{F}(x) = 1 - F(x)$. The case $h = k$ is trivial. For $h < k$ using the Markovian property of order statistics, we can write for any g that

$$\mathbf{E}\big[g\left(X_{(h+1)}^*\right)|X_{(h)}^* = x\big] = \frac{-1}{\bar{F}^{\beta_h}(x)} \int_x^\infty g(u)\bar{F}^{\beta_h}(du) = 0.$$

Introduce the sets $B_{(h+1)} = \{x : g(x) > 0\}$ and $B_- = \{x : g(x) < 0\}$. Represent the integral as the difference of two singular measures

$$\int_x^\infty g(x)\bar{F}^{\beta_h}(dx) = \mu_+((x, \infty)) - \mu_-((x, \infty)),$$

where $\mu_\pm(A) = \int_{A \cap B_\pm} g(u)\bar{F}^{\beta_h}(du)$ for any Borel set A. Hence, the measures μ_+ and μ_- coincide on any semi-interval (x, ∞). Therefore, $\mu_+(A) = \mu_-(A)$ for any Borel set A. Now the condition that they are singular leads to $\mu_+(A) = \mu_-(A) = 0$ for any Borel set A. Taking account of the equivalence of the measure generated by $X_{(h+1)}^*$ and the measure under the integral yields that $g(X_{(h+1)}^*) = 0$ a.s. Using the same operation for $h = h+1, \ldots, k-1$, one obtains Condition (6.2.1) for any measurable integrable function g. Now we choose

$$g(X_{(k)}^*) = \mathbf{E}[f(Z)|X_{(k)}^*] - \mathbf{E}[f(Z)].$$

The statement follows directly from Lemma 6.2.1. ∎

REMARK 6.2.1 The property that the independence of Z and $X_{(h)}^*$ implies the independence of Z and $X_{(k)}^*$ and also with the vector $X_{(1)}^*, \ldots, X_{(k)}^*$ follows immediately from the lemma. For this situation, a similar result was given independently for generalized order statistics by Keseling and Kamps (2002).

To obtain a similar result with the corresponding spacings instead of the usual order statistics, Rossberg (1972) used integrated-type properties that if $g(x) = \mathbf{P}(Z < y|X_{(h)} = x) - \mathbf{P}(Z < y)$, then the condition

$$\int_x^\infty \int_{-\infty}^\infty g(u) F_{X_{(h)}|\Delta_{(h)}}(du|s) \, dF_{\Delta_h}(s)$$
$$= \int_{-\infty}^\infty \int_x^\infty g(u) F_{\Delta_h|X_{(h)}}(ds|u) F_{X_{(h)}}(du) = 0,$$

or (equivalently)

$$\int_{-\infty}^\infty g(s)\mathbf{P}(\Delta_h > x|X_{(h)} = s) \, dF_{X_{(h)}}(s) = 0, \qquad (6.2.2)$$

for dF_{Δ_h}-almost all x, implies that $g(X_{(h)}) = 0$ a.s. In general, this is not true. For example, under the condition $F(x) = \exp(x)\mathbb{I}_{\{x<0\}} + \mathbb{I}_{\{x \geq 0\}}$, the random variables $X_{(h)}$ and Δ_h are independent. Thus, we can take any function g such that $\mathbf{E}[g(X_{(h)})] = 0$. Rossberg used the condition that $x_0 = \inf\{x : F(x) > 0\} > -\infty$ and the Laplace transform is non-zero for all $z \in E = E_+ \cup D$, where $E_+ = \{x \in \mathbb{C} : \mathrm{Re}(x) > 0\}$, $D = \{x \in \mathbb{C} : \mathrm{Re}(x) = 0\}$.

LEMMA 6.2.3 *Let F be a distribution function such that $x_0 = \inf\{x : F(x) > 0\} = 0$ and the Laplace transform is non-zero for all $z \in E = E_+ \cup D$, and R be some function of bounded variation such that $R(x) = 0$ for all $x \leq 0$. Then the condition*

$$\int_0^\infty F(x-t) \, dR(x) = 0, \qquad \text{for any } t > 0$$

implies that $R(x) = 0$ for almost all $x \geq 0$.

We now formulate the following lemma in the case of Type-II progressively censored order statistics.

LEMMA 6.2.4 *Let $X_1, \ldots, X_n$ be i.i.d. random variables having a distribution function F and $x_0 = \inf\{x : F(x) > 0\} > -\infty$. Further, suppose the Laplace transform*

$$F^*(z) = -\int_0^\infty e^{-zx} d\bar{F}_h(x + x_0)$$

has no zeroes in E, where

$$F_h(t) = \sum_{i=1}^{h-1} \frac{a_{i,h-1}}{\gamma_{i,h-1}} \{1 - F(t)\}^{\gamma_{i,h-1}},$$

$$a_{i,r} = \prod_{\substack{j=1 \\ j \neq i}}^{r} \frac{1}{\beta_{j-1} - \beta_{i-1}},$$

and $\gamma_{i,r-1} = n - m - \beta_{r-1} + \beta_{i-1} + 1$. Suppose also $X_{(1)}^*(h), \ldots,$ $X_{(m-h)}^*(h)$ are generalized Type-II progressively censored order statistics from the sample $X_{\alpha_h+1}, \ldots, X_n$ with $R_{h+1}, \ldots, R_m$ and

$$\mathbf{E} \mid f \left(\phi(X_{(k-h)}^*(h), \ldots, X_{(m-h)}^*(h)) \right) \mid < \infty. \tag{6.2.3}$$

Then for any k and h such that $1 \leq h < k < m$, the condition $\mathbf{E}[f(Z)|X_{(h)}^*] = c = constant$ implies that $\mathbf{E}[f(Z)|X_{(k)}^*] = \mathbf{E}[f(Z)|X_{(1)}^*, \ldots, X_{(k)}^*] = c$.

PROOF. Without loss of any generality, let us assume that $x_0 = 0$. Note that

$$\mathbf{P}(\Delta_h^* > v|X_{(h)}^* = x)$$

$$= \frac{\beta_{h-1} \{1 - F(x)\}^{\beta_{h-1}-1} \sum_{i=1}^{h-1} \frac{a_{i,h-1}}{\gamma_{i,h-1}} \{1 - (1 - F(x-v))^{\gamma_{i,h-1}}\}}{(n - m + \beta_{h-1}) \sum_{i=1}^{h} a_{i,h} \{1 - F(x)\}^{n-m+\beta_{i-1}-1}}$$

for all $v > 0$, and

$$\mathbf{P}(X_{(h)}^* < x) = 1 - C_{h-1} \sum_{i=1}^{h} \frac{a_{i,h}}{n - m + \beta_{i-1}} \{1 - F(x)\}^{n-m+\beta_{i-1}},$$

where $C_{h-1} = \prod_{i=0}^{h-1}(n-m+\beta_i)$; see Balakrishnan, Cramer, and Kamps (2001). Then, after substituting in (6.2.2), we obtain

$$\int_0^\infty \sum_{i=1}^h \frac{a_{i,h-1}}{\gamma_{i,h-1}} \left\{1-(1-F(x-v))^{\gamma_{i,h-1}}\right\} g(x)\, dR(x) = 0$$

for all $v > 0$, where $R(x) = C_{h-2}\left\{1-(1-F(x))^{\beta_{h-1}}\right\}$ is the continuous non-decreasing function such that the finite measure generated by R on line is equivalent to the measure generated by F. We choose

$$g(x) = \mathbf{E}[f(Z)|X_{(k)}^* = x] - \mathbf{E}\,[f(Z)].$$

Then the condition

$$\left|\int_{-\infty}^\infty \int_x^\infty \mathbf{E}[f(Z)|X_{(h)}^* = u] F_{\Delta_h^*|X_{(h)}^*}(ds|u) F_{X_{(h)}^*}(du)\right|$$
$$= |\mathbf{E}[f(Z)\mathbb{1}_{\{\Delta_h^* > x\}}]|$$
$$\leq \mathbf{E}|f(Z)| < \infty$$

for all $x > 0$ and the condition (6.2.3)

$$\mathbf{E}[|f(Z)|\,|X_{(h)}^* = 0]$$
$$= \mathbf{E}\,|f\left(\phi(X_{(k-\alpha_h)}^*(h),\dots,X_{(n-\alpha_h)}^*(h))\right)| < \infty$$

leads to the bounded variation condition for $\int_0^x g(u)\,R(du)$. Therefore, by Lemma 6.2.3 we can conclude that $g(X_{(h)}^*) = 0$ a.s. Then the result follows readily from Lemma 6.2.2. ∎

The next lemma is a modification of the characterization theorem of Rossberg (1972). A similar characterization has been given independently for any generalized order statistics in the Ph.D. thesis of Claudia Keseling (1999).

LEMMA 6.2.5 *Let $X_1,\dots,X_n$ be i.i.d. random variables having a continuous distribution function F. Then $F(x) = F^*(ax+b)$, $x \in \mathbb{R}$, for some $a, b \in \mathbb{R}$, where F^* is the standard exponential distribution function, iff $X_{(k)}^*$ is independent of some linear combination $\sum_{j=k}^l s_j X_{(j)}^*$, such that $\sum_{j=k}^l s_j = 0$, $s_k \neq 0$ and $s_l \neq 0$, $k \leq l \leq m$.*

SKETCH OF PROOF. Using Representation 6.2.2 and continuity of the distribution function F, we can write [following Rossberg (1972)] that

$$\mathbf{E} \exp[z\{s_k G(y) + s_{k+1}G(y + \delta_{k+1}) + \cdots$$
$$+ s_l G(y + \delta_{k+1} \ldots + \delta_l)\}]$$
$$= C_{l,k} \int_0^\infty \cdots \int_0^\infty \exp[z\{s_k G(y) + s_{k+1}G(y + u_1) + \cdots$$
$$+ s_l G(y + u_1 \ldots + u_{l-k})\}]$$
$$\times \exp\left(-\sum_{j=1}^{l-k} \beta_{k+j-1} u_j\right) du_1 \ldots du_{l-k}$$
$$= K(z) = constant.$$

Let us introduce the following notations

$$a_i(z, v) = \exp\{z s_{k+i} G(v) - (r_{k+i} + 1)v\},$$
$$b_i(z, y) = \beta_{i+k-1} y + z d_i G(y), \quad b_i'(z, y) = \frac{\partial b_i(z, y)}{\partial y},$$
$$P_1(y, z) = b_i'(z, y), \quad \text{and}$$
$$P_i(y, z) = \frac{\partial}{\partial y} P_{i-1}(y, z) - b_i'(z, y) P_{i-1}(z, y),$$

where $d_i = \sum_{j=1}^i s_{k+j-1}$. By setting $v_i = y + u_1 + \cdots + u_i$, the previous equality can be rewritten in the form

$$\exp\{b_1(z, y)\} \int_y^\infty dv_1 a_1(z, v_1) \ldots \int_{v_{l-k-2}}^\infty dv_{l-k-1} a_{l-k-1}(z, v_{l-k-1})$$

$$\times \int_{v_{l-k-1}}^\infty dv_{l-k} \exp\{-b_{l-k}(z, v_{l-k})\} = K(z).$$

Under the condition $l - k = 1$, we can write it as

$$C_{l,k} \int_0^\infty e^{z\{s_{k+1}G(y+u) + s_k G(y)\}} e^{-\beta_k u} du$$
$$= C_{l,k} e^{\beta_k y + z s_k G(y)} \int_y^\infty e^{z s_{k+1} G(v) - \beta_k v} dv = K(z),$$

which leads to the equality

$$C_{l,k}e^{zs_{k+1}G(y)-\beta_k y}=\{\beta_k+zG'(y)\}e^{-zs_k G(y)-\beta_k y}K(z).$$

Taking account of the condition $s_{k+1}=-s_k$, we obtain

$$\{\beta_k+zG'(y)\}K(z)=P_1(z,y)\,K(z)=C_{l,k}$$

for all $y\in\mathbb{R}$. Therefore, $G'(y)=a=constant$ and $G(y)=ay+b$ for some $a,b\in\mathbb{R}$. Let us mention that the difference equation used by Rossberg (1972)

$$a_i(z,y)\exp\{b_i(z,y)\}=\exp\{b_{i+1}(z,y)\},$$

for $i\in\mathbb{N}$, is also valid. Thus, we can also use here the process used by Rossberg (1972). ∎

We can also modify the characterization theorem for the exponential distribution via some regression properties of order statistics; see Rao and Shanbhag (1994).

LEMMA 6.2.6 *Let $X_1,\ldots,X_n$ be i.i.d. random variables having a continuous distribution function F. Then $F(x)=F^*(ax+b)$, $x\in\mathbb{R}$, for some $a\in\mathbb{R}$ and some fixed $b\in\mathbb{R}$, iff for some non-arithmetic (non-lattice) monotonic real function f, $\mathbf{E}[f(\Delta_k^*)|X_{(k)}^*]=c$ for some fixed $c\neq f(0+)$.*

SKETCH OF PROOF. We can use the same arguments as in Rao and Shanbhag (1994), changing i by α_i. ∎

Using Lemmas 6.2.2 and 6.2.4–6.2.6, we obtain the following characterization theorems.

THEOREM 6.2.1 *Let $X_1,\ldots,X_n$ be i.i.d. random variables having a continuous distribution function F, $1\le h\le k<m$. Then the following conditions are equivalent:*

(i) $F(x)=F^*(ax+b)$, $x\in\mathbb{R}$, for some $a,b\in\mathbb{R}$;

(ii) $X_{(h)}^*$ is independent of some non-trivial linear combination

$$\sum_{j=1}^{m-k}c_j\Delta_{k+j}^*=\sum_{j=0}^{m-k}s_j X_{(k+j)}^*,$$

with $\sum_{j=0}^{m-k}s_j=0$;

(iii) *The conditions of Lemma 6.2.4 hold and $\Delta_{(h)}^*$ is in-dependent of some non-trivial linear combination*

$$\sum_{j=1}^{m-k} c_j \Delta_{k+j}^* = \sum_{j=0}^{m-k} s_j X_{(k+j)}^*,$$

with $\sum_{j=0}^{m-k} s_j = 0$.

THEOREM 6.2.2 *Let $X_1, \ldots, X_n$ be i.i.d. random variables having a continuous distribution function F and let f be a non-arithmetic monotone real function; let c be a constant such that $c \neq f(0)$, $1 \leq h < k \leq m$. Then the following conditions are equivalent:*

(i) $F(x) = F^*(ax + b)$, $x \in \mathbb{R}$, *for some $a \in \mathbb{R}$ and a fixed $b \in \mathbb{R}$;*

(ii) $\mathbf{E}[f(\Delta_k^*)|X_{(h)}^*] = c$;

(iii) *The conditions of Lemma 6.2.4 hold and $\mathbf{E}(f(\Delta_k^*)| \Delta_h^*) = c$.*

In the dependent case, the situation is more complicated. As an example, we give the following characterization in the class of mixed i.i.d. exponential distributions. Let (X_1, X_2) have an exchangeable absolutely continuous distribution with a density function

$$p(x, y) = \int_\Omega \alpha^2(\omega) e^{-\alpha(\omega)(x+y)} Q(d\omega) = G(x+y),$$

where Q is some probability measure on $(\Omega, \mathfrak{F})$. In this case by the exchangeability property, the joint density of maximum and minimum have the form

$$p_{X_{(2)}, X_{(1)}}(x, y) = 2p(x, y) \mathbb{1}_{\{x \geq y\}},$$

from which we obtain the following:

$$p_{\Delta_2, X_{(1)}}(t, y) = 2p(t+y, y) \mathbb{1}_{\{t \geq 0\}},$$

$$p_{\Delta_2}(x) = \int_\Omega \alpha(\omega) e^{-\alpha(\omega)x} Q(d\omega) = H(x),$$

$$p_{X_{(1)}}(x) = \int_\Omega 2\alpha(\omega) e^{-2\alpha(\omega)x} Q(d\theta).$$

The independence of $X_{(2)} - X_{(1)}$ and $X_{(1)}$ then leads to the equality

$$2 \int_{\Omega} \alpha^2(\omega) e^{-\alpha(\omega)(t+2y)} Q(d\omega) = 2 \int_{\Omega} \alpha(\omega) e^{-\alpha(\omega)t} Q(d\omega)$$
$$\times \int_{\Omega} \alpha(\omega) e^{-\alpha(\omega)(2y)} Q(d\omega),$$

or

$$G(t+s) = H(t)H(s)$$

for all $t, s > 0$. This equality characterizes the exponential function with a constant, i.e., $H(x) = ae^{bx}$. For continuous H we can write $G(t) = H(t)H(0)$. This leads to the well-known equality

$$R(t+s) = R(t)R(s),$$

where $R(t) = H(t)/H(0)$, which characterizes the exponential function. In our situation, $H(t)$ must be a density function. Therefore, $H(t) = ae^{-\alpha t}$ for all $t > 0$, which yields the following characterization result.

THEOREM 6.2.3 *Let (X_1, X_2) have an exchangeable absolutely continuous distribution with a joint density function*

$$p(x, y) = \int_{\Omega} \alpha^2(\omega) e^{-\alpha(\omega)(x+y)} Q(d\omega),$$

where Q is some probability measure on $(\Omega, \mathfrak{F})$. Then $X_{(1)}$ and $X_{(2)} - X_{(1)}$ are independent iff $p(x, y) = \alpha^2 e^{-\alpha(x+y)} \mathbb{I}_{\{x \geq 0, \, y \geq 0\}}$ for some $\alpha > 0$, i.e., X_1 and X_2 are independent random variables.

REFERENCES

Balakrishnan, N. and Aggarwala, R. (2000). *Progressive Censoring: Theory, Methods and Applications*. Birkhäuser, Boston, Massachusetts.

Balakrishnan, N., Cramer, E. and Kamps, U. (2001). Bounds for means and variances of progressive type II censored order statistics. *Statistics & Probability Letters*, **54**, 301–315.

Beg, M. I. and Kirmani, S. N. U. A. (1979). On characterizing the exponential distribution by a property of truncated spacings. *Sankhyā, Series A*, **41**, 278–284.

Basu, D. (1955). On statistics independent of sufficient statistics. *Sankhyā, Series A*, **15**, 377–380.

Fisz, M. (1958). Characterization of some probability distributions. *Skandinavisk Aktuarietidskrift*, **41**, 65–70.

Kamps, U. (1995). *A Concept of Generalized Order Statistics*. Teubner, Stuttgart, Germany.

Keseling, C. and Kamps, U. (2002). A theorem of Rossberg for generalized order statistics. *Sankhyā* **65**, 259–270.

Keseling, C. (1999). *Characterizations of Probability Distributions by Generalized Order Statistics (in German)*. Ph.D. Thesis, Aachen University of Technology, Aachen, Germany.

Lau, K. S. and Rao, C. R. (1982). Integrated Cauchy functional equation and characterizations of the exponential law. *Sankhyā, Series A*, **44**, 72–90.

Rao, C. R. and Shanbhag, D. N. (1994). *Choquet-Deny Type Functional Equations with Applications to Characterizations*. John Wiley & Sons, Chichester, U.K.

Rogers, G. (1959). A note on the stochastic independence of functions of order statistics. *Annals of Mathematical Statistics*, **30**, 1263–1264.

Rogers, G. (1963). An alternative proof of the characterization of the density. *American Mathematical Monthly*, **70**, 857–858.

Tanis, E. A. (1964). Linear forms in the order statistics from an exponential distribution. *Annals of Mathematical Statistics*, **35**, 270–276.

Chapter 7

A Note on Regressing Order Statistics and Record Values

I. BAIRAMOV
Department of Mathematics, İzmir University of Economics, İzmir, Turkey

N. BALAKRISHNAN
Department of Mathematics and Statistics, McMaster University, Hamilton, Ontario, Canada

CONTENTS

ABSTRACT

Let $X_1, X_2, \ldots, X_n$ be independent and identically distributed random variables with a common distribution function F. Let $X_{1:n} \leq X_{2:n} \leq \ldots \leq X_{n:n}$ denote the order statistics constructed from $X_1, X_2, \ldots, X_n$. The regression of order statistics, i.e., $E(X_{k+m:n} \mid X_{k:n} = x) \equiv \Psi_{k,m:n}(x)$, $k \geq 2$, $m + k \leq n$, is considered. An inverse formula expressing F through $\Psi_{k,m:n}(x)$

and $\Psi_{k,m-1:n-1}(x)$ is given. Specifically, it is shown that knowledge of $\Psi_{k,m:n}(x)$ and $\Psi_{k-1,m:n-1}(x)$, for some $n \geq m + k$, $k \geq 1$, $m \geq 1$, amounts to knowing the distribution F. Similarly, the regression of record values is also considered, and a parallel result is established.

KEYWORDS AND PHRASES: Order statistics, record values, regression, characterization of distributions

7.1   INTRODUCTION

Let $X_1, X_2, \ldots, X_n$ be independent and identically distributed (i.i.d.) random variables with a common distribution function (d.f.) F. Let $X_{1:n} \leq X_{2:n} \leq \ldots \leq X_{n:n}$ denote the order statistics constructed from $X_1, X_2, \ldots, X_n$. With respect to the squared-error loss, the best unbiased predictor for $X_{m+k:n}$, given $X_{k:n}$, is $E(X_{m+k:n} \mid X_{k:n})$. Ferguson (1967) was the first author to consider the problem of determining all d.f.'s F for which

$$E(X_{k+m:n} \mid X_{k:n}) = cX_{k:n} + d \text{ a.s.}$$

and provided a complete solution for $m = 1$. In general, the problem can be stated as follows. Let

$$E(G(X_{1:n}, X_{2:n}, \ldots, X_{n:n}) \mid X_{k:n}) = H(X_{k:n}) \text{ a.s.,} \qquad (7.1.1)$$

where $G : R^n \to R$ and $H : R \to R$ are known functions and $k \in \{1, 2, \ldots, n\}$ is fixed. The objective then is to find all d.f.'s F for which (7.1.1) holds.

Wesolowski and Ahsanullah (1997) noted that the general solution of this problem is not known and listed several papers which have dealt with special cases. In the paper of Wesolowski and Ahsanullah (1997), the solution for the case of absolutely continuous distributions with $m = 2$, $H(x) = cx+d$, $G(x_1, x_2, \ldots, x_n) = x_{k+2}$ is given. The main result of their paper may be expressed as follows. Let $E(|X_{k+2:n}|) < \infty$. If the regression of $X_{k+2:n}$ on $X_{k:n}$ is linear, i.e., $E(X_{k+2:n} \mid X_{k:n}) = cX_{k:n}+d$, then X has a Pareto distribution if $c > 1$, a Power distribution if $c < 1$, and an exponential distribution if $c = 1$.

Blazquez and Rebollo (1997) obtained a solution for re-gression $X_{k+m:n}$ on $X_{k:n}$, $1 \le k < k+m \le n$. Let $D_F = \{x : 0 < F(x) < 1\}$. Their result is as follows:

Let X be a r.v. with d.f. F which is k times differentiable in D_F, such that

$$E(X_{k+m:n} \mid X_{k:n}) = \beta X_{k:n} + \alpha.$$

Then, except for location and scale parameters,

$$
\begin{aligned}
F(x) &= 1 - |x|^\delta, & \text{for } x \in [-1, 0], && \text{if } 0 < \beta < 1 \\
&= 1 - \exp(-x), & \text{for } x \in [0, \infty], && \text{if } \beta = 1 \\
&= 1 - x^\delta, & \text{for } x \in [1, \infty], && \text{if } \beta > 1,
\end{aligned}
$$

where $\delta = (r - (n-m))^{-1}$ and r is the unique real root greater than $m - 1$ of the polynomial equation

$$P_m(z) = \frac{1}{\beta} P_m(n - k),$$

$$P_k(z) = z(z - 1) \dots (z - k + 1).$$

Another interesting result is due to Dembinska and Wesolowski (1998), who consider the linearity of regression $E(X_{k:n} \mid X_{k+r:n}) = cX_{k+r:n} + d$ for some $k \le n - r$ and some real c and d in the case of continuous underlying distribu-tion function. They prove that only three cases are possible: the negative exponential distribution, negative Pareto distri-bution, and negative Power distribution.

In this paper, however, we consider the general case for $k \ge 1$, $m + k \le n$, and

$$E(X_{k+m:n} \mid X_{k:n} = x) \equiv \Psi_{k,m:n}(x). \tag{7.1.2}$$

Assuming that X has an absolutely continuous d.f. F, we obtain an inverse formula. Specifically, we show that knowl-edge of $\Psi_{k,m:n}(x)$ and $\Psi_{k,m-1:n-1}(x)$, for some n, m, k such that $n \ge m+k$, $k \ge 2$, $m \ge 1$, where $\Psi_{k,m-1:n-1}(x) \equiv E(X_{k+m-1:n-1} \mid X_{k:n-1} = x)$ amounts to knowing the distribution F.

Let $[a, b]$ be the support of the X, $-\infty \le a < b \le \infty$. Then Theorem 7.2.1 shows that if X is an absolutely con-tinuous r.v. with d.f. F, under an integrability assumption, for any $x \in (a, b)$

$$\bar{F}(x) = \bar{F}(a) \exp\left\{ -\frac{1}{n-k} \int_a^x \frac{\Psi'_{k,m:n}(t)}{\Psi_{k,m:n}(t) - \Psi_{k,m-1:n-1}(t)} dt \right\},$$

where $\Psi'_{k,m:n}(t) = \frac{d}{dt}\Psi_{k,m:n}(t)$ and $\bar{F}(x) = 1 - F(x)$.

Theorem 7.2.2, a special case of Theorem 7.2.1, shows that with $\varphi_n(x) = E(X_{n:n} \mid X_{1:n} = x)$, we have

$$\bar{F}(x) = \bar{F}(a) \exp\left\{ -\frac{1}{n-1} \int_a^x \frac{\varphi'_n(t)}{\varphi_n(t) - \varphi_{n-1}(t)} dt \right\}.$$

For a sequence of i.i.d. r.v.'s $X_1, X_2, \ldots, X_n$ with a common continuous d.f. F, we define the record times of the sequence as follows:

$$U(1) = 1,$$

$$U(n) = \min\left\{ i : i > U(n-1), X_i > X_{U(n-1)} \right\}, \quad n = 2, 3, \ldots$$

Let $X_{U(1)}, X_{U(2)}, \ldots, X_{U(n)}, \ldots$ be corresponding record values. For detailed reviews on record values and their applications, one may refer to Nevzorov (1987); Nagaraja (1988); Arnold, Balakrishnan, and Nagaraja (1992, 1998); Ahsanullah (1995); and Nevzorov and Balakrishnan (1998).

Recently, Dembinska and Wesolowski (2000) considered the problem of identifying all the distributions allowing linearity of regressions either $E(X_{U(m+k)} \mid X_{U(m)})$ or $E(X_{U(m)} \mid X_{U(m+k)})$. Their result for $E(X_{U(m+k)} \mid X_{U(m)})$ is similar to the result of Blazquez and Rebollo (1997) possessing exponential, Pareto, and Power distributions (up to location and scale parameters) but obtained without some stringent smoothness assumptions on the d.f. F. Note that all previous results cited above were obtained by solving differential equations. Dembinska and Wesolowski used an extended version of integrated Cauchy functional equations [see Rao and Shanbag (1994)].

We consider the general case when $E(X_{U(m+k)} \mid X_{U(m)})$ is not necessarily linear. Let

$$\varphi_{n,k}(x) = E(X_{U(n+k)} \mid X_{U(n)} = x).$$

Then Theorem 7.3.1 establishes that

$$\bar{F}(x) = \bar{F}(a) \exp\left\{-\frac{1}{n-1} \int_a^x \frac{\varphi'_{n,k}(t)}{\varphi_{n,k}(t) - \varphi_{n,k-1}(t)} dt\right\}.$$

7.2 REGRESSING ORDER STATISTICS

For the i.i.d. r.v.'s $X_1, X_2, \ldots, X_n$ with absolutely continuous common d.f. F, consider (7.1.2) for $m + k \le n$

$$\Psi_{k,m:n}(x) \equiv E(X_{k+m:n} \mid X_{k:n} = x).$$

The probability density function of $X_{k:n}$ is

$$f_{k:n}(x) = \frac{n!}{(k-1)!(n-k)!} F^{k-1}(x)\{1 - F(x)\}^{n-k} f(x), \quad (7.2.1)$$

where $f(x) = F'(x)$. The joint density function of $X_{k:n}$ and $X_{k+m:n}$ is [see Arnold, Balakrishnan, and Nagaraja (1992, p. 16)]

$$f_{k,k+m:n}(x, y)$$

$$= \frac{n!}{(k-1)!(m-1)!(n-k-m)!} F^{k-1}(x)$$

$$\times \{F(y) - F(x)\}^{m-1} \times \{1 - F(y)\}^{n-k-m} f(x) f(y), \quad \text{if } x < y$$

$$= 0 \quad \text{otherwise.} \tag{7.2.2}$$

Using (7.2.1) and (7.2.2), we obtain

$$\Psi_{k,m:n}(x) = \frac{1}{f_{k:n}(x)} \int_x^b y f_{k,k+m:n}(x, y) \, dy$$

$$= \frac{(k-1)!(n-k)!}{n!} \frac{1}{F^{k-1}(x)\{1 - F(x)\}^{n-k} f(x)}$$

$$\times \frac{n!}{(k-1)!(m-1)!(n-k-m)!} \int_x^b y F^{k-1}(x)$$

$$\times \{F(y) - F(x)\}^{m-1}\{1 - F(y)\}^{n-k-m} f(x) f(y) \, dy$$

$$= m C_{n-k}^m \frac{1}{\{\bar{F}(x)\}^{n-k}} \int_x^b y\{F(y) - F(x)\}^{m-1}$$

$$\times \{1 - F(y)\}^{n-k-m} f(y) \, dy, \tag{7.2.3}$$

where, $C_n^m = \frac{n!}{m!(n-m)!}$. Similarly, we have

$$\Psi_{k,m-1:n-1}(x)$$

$$= (m-1)C_{n-k-1}^{m-1}\frac{1}{\{\bar{F}(x)\}^{n-1-k}}\int_x^b y\{F(y)-F(x)\}^{m-2}$$

$$\times\{1-F(y)\}^{n-k-m}f(y)\,dy \quad \text{for all } x \in (a,b). \quad (7.2.4)$$

THEOREM 7.2.1 *Let $X_1, X_2, \ldots, X_n$ be i.i.d. r.v.'s with an absolutely continuous d.f. F. Let $[a,b]$ be the support of the X, $-\infty \le a < b \le \infty$. Then, under an integrability assumption, for $n \ge k+m$, $k \ge 2$, $m \ge 1$, it is true that for any $x \in (a,b)$*

$$\bar{F}(x) = \bar{F}(a)\exp\left\{-\frac{1}{n-k}\int_a^x \frac{\Psi'_{k,m:n}(t)}{\Psi_{k,m:n}(t)-\Psi_{k,m-1:n-1}(t)}\,dt\right\},$$

and for $m = 1$, $n \ge k+1$,

$$\bar{F}(x) = \bar{F}(a)\exp\left\{-\frac{1}{n-m}\int_a^x \frac{\Psi'_{k:n}(t)}{\Psi_{k:n}(t)-t}\,dt\right\},$$

where $\Psi_{k:n}(t) = \Psi_{k,1:n}(t)$.

PROOF. Rewriting (7.2.3) as

$$\Psi_{k,m:n}(x)(\bar{F}(x))^{n-k} = mC_{n-k}^m\int_x^b y\{F(y)-F(x)\}^{m-1}$$

$$\times\{1-F(y)\}^{n-k-m}f(y)\,dy \quad (7.2.5)$$

and differentiating (7.2.5) with respect to x, we obtain

$$\Psi'_{k,m:n}(x)\{\bar{F}(x)\}^{n-k} - (n-k)\{\bar{F}(x)\}^{n-k-1}f(x)\Psi_{k,m:n}(x)$$

$$= -m(m-1)C_{n-k}^m f(x)\int_x^b y\{F(y)-F(x)\}^{m-2}$$

$$\times\{1-F(y)\}^{n-m-k}f(y)\,dy - mC_{n-k}^m y\{F(y)-F(x)\}^{m-1}$$

$$\times\{1-F(y)\}^{n-k-m}f(y)\,|_{x=y}$$

$$= -m(m-1)C_{n-k}^m f(x)\int_x^b y\{F(y)-F(x)\}^{m-2}$$

$$\times\{1-F(y)\}^{n-m-k}f(y)\,dy. \quad (7.2.6)$$

Consider now the right-hand side of (7.2.6). Noting first that

$$m(m-1)C^m_{n-k} = (n-k)(m-1)C^{m-1}_{n-1-k}$$

and then using (7.2.4) in (7.2.6), we get

$$\Psi'_{k,m:n}(x)\{\bar{F}(x)\}^{n-k} - (n-k)$$

$$\times\{\bar{F}(x)\}^{n-k-1} f(x)\Psi_{k,m:n}(x)\{\bar{F}(x)\}^{n-k}$$

$$= -(n-k)\Psi_{k,m-1:n-1}(x)\{\bar{F}(x)\}^{n-k-1} f(x)$$

so that

$$\frac{f(x)}{\bar{F}(x)} = \frac{1}{n-k}\frac{\Psi'_{k,m:n}(x)}{\Psi_{k,m:n}(x) - \Psi_{k,m-1:n-1}(x)}. \tag{7.2.7}$$

Integrating (7.2.7) in $[a,t]$, we obtain the assertions of the theorem. ∎

The following theorem is a special case of Theorem 7.2.1.

THEOREM 7.2.2 *Let*

$$\varphi_n(x) = E(X_{n:n} \mid X_{1:n} = x).$$

Then, under the assumptions of Theorem 7.2.1, we have

$$\bar{F}(x) = \bar{F}(a)\exp\left\{-\frac{1}{n-1}\int_a^x \frac{\varphi'_n(t)}{\varphi_n(t) - \varphi_{n-1}(t)}\,dt\right\}.$$

EXAMPLE 7.2.1 *Let* $\Psi_{k,m:n}(x) = \alpha_1 t + \beta_1$ *and* $\Psi_{k,m-1:n-1}(t) = \alpha_2 t + \beta_2$, *where* α_i, β_i $(i = 1,2)$ *are some constants. Let* $\alpha = \alpha_1 - \alpha_2$ *and* $\beta = \beta_1 - \beta_2$. *Then by Theorem 7.2.1, we have*

$$F(x) = 1 - \exp\left\{-\frac{1}{n-k}\int_a^x \frac{\alpha_1}{\alpha t + \beta}\,dt\right\}$$

$$= 1 - \left(\frac{\alpha x + \beta}{\alpha a + \beta}\right)^{-\frac{\alpha_1}{\alpha(n-k)}}, \quad a < x < \infty,$$

which is a translated Pareto distribution.

7.3 REGRESSING RECORD VALUES

Let $X_1, X_2, \ldots, X_n, \ldots$ be a sequence of i.i.d. r.v.'s with an absolutely continuous d.f. F and density function f. Consider

the record values of this sequence $X_{U(1)}, X_{U(2)}, \ldots, X_{U(n)}, \ldots$.
Denote

$$\varphi_{n,k}(x) = E(X_{U(n+k)} \mid X_{U(n)} = x). \tag{7.3.1}$$

The joint density of $X_{U(n)}$ and $X_{U(m)}$ (for $m > n$) is [see Arnold, Balakrishnan, and Nagaraja (1998, p.11)]

$$f_{m,n}(x, y) = \frac{\{R(x)\}^{n-1}}{(n-1)!} r(x) \frac{\{R(y) - R(x)\}^{m-n-1}}{(m-n-1)!} f(y),$$

$$\text{for } x < y, \tag{7.3.2}$$

where $R(x) = -\ln\{1 - F(x)\}$ and $r(x) = \frac{d}{dx} R(x) = \frac{f(x)}{1-F(x)}$. From Equations (7.3.1) and (7.3.2), we have

$$\varphi_{n,k}(x) = \frac{1}{(k-1)!} \frac{1}{\bar{F}(x)} \int_x^b y(\ln \bar{F}(x) - \ln \bar{F}(y))^{k-1} f(y) \, dy. \tag{7.3.3}$$

It can be seen from (7.3.3) that

$$E(X_{U(n+k)} \mid X_{U(n)} = x) = E(X_{U(1+k)} \mid X_1 = x)$$

for any n. Then

$$\varphi_{n,k-1}(x) = E(X_{U(k)} \mid X_1 = x)$$

$$= \frac{1}{(k-2)!} \frac{1}{\bar{F}(x)} \int_x^b y\{\ln \bar{F}(x) - \ln \bar{F}(y)\}^{k-2} f(y) \, dy. \tag{7.3.4}$$

Then we have the following result.

THEOREM 7.3.1 *Under integrability assumptions, it is true that for any $x \in (a, b)$*

$$\bar{F}(x) = \bar{F}(a) \exp\left\{ -\frac{1}{n-1} \int_a^x \frac{\varphi'_{n,k}(t)}{\varphi_{n,k}(t) - \varphi_{n,k-1}(t)} \, dt \right\}.$$

PROOF. From (7.3.3), we have

$$\varphi_{n,k}(x)\bar{F}(x) = \frac{1}{(k-1)!} \int_x^b y\{\ln \bar{F}(x) - \ln \bar{F}(y)\}^{k-1} f(y) \, dy. \tag{7.3.5}$$

Differentiating (7.3.5) with respect to x and then using (7.3.4), we obtain

$$\varphi'_{n,k}(x)\bar{F}(x) - f(x)\varphi_{n,k}(x)$$

$$= -\frac{k-1}{(k-1)!}\int_x^b y\{\ln \bar{F}(x) - \ln \bar{F}(y)\}^{k-2}f(y)\frac{f(x)}{\bar{F}(x)}\,dy$$

$$\quad -\frac{1}{(k-1)!}y\{\ln \bar{F}(x) - \ln \bar{F}(y)\}^{k-1}f(y)\mid_{y=x}$$

$$= -\frac{1}{(k-2)!}\frac{f(x)}{\bar{F}(x)}\int_x^b y\{\ln \bar{F}(x) - \ln \bar{F}(y)\}^{k-2}f(y)\,dy$$

$$= -f(x)\varphi_{n,k-1}(x),$$

Hence,

$$\frac{f(x)}{\bar{F}(x)} = \frac{\varphi'_{n,k}(x)}{\varphi_{n,k}(x) - \varphi_{n,k-1}(x)}. \tag{7.3.6}$$

Integrating (7.3.6) in $[a, t]$, we obtain the result stated in the theorem. ∎

REFERENCES

Ahsanullah, M. (1995). *Record Statistics*. Nova Science Publishers, Commack, New York.

Arnold, B. C., Balakrishnan, N. and Nagaraja, H. N. (1992). *A First Course in Order Statistics*. John Wiley & Sons, New York.

Arnold, B. C., Balakrishnan, N. and Nagaraja, H. N. (1998). *Records*. John Wiley & Sons, New York.

Blazquez, F. L. and Rebollo, J. L. M. (1997). A characterization of distributions based on linear regression of order statistics and record values. *Sankhyā, Series A*, **59**, 311–323.

Dembinska, A. and Wesolowski, J. (1998). Linearity of regression for non-adjacent order statistics. *Metrika*, **48**, 215–278.

Dembinska, A. and Wesolowski, J. (2000). Linearity of regression for non-adjacent record values. *Journal of Statistical Planning and Inference*, **90**, 195–205.

Ferguson, T. S. (1967). On characterizing distributions by properties of order statistics. *Sankhyā, Series A*, **29**, 265–278.

Nagaraja, H. N. (1988). Record values and related statistics—A review. *Communications in Statistics—Theory and Methods*, **17**, 2223–2238.

Nevzorov, V. B. (1987). Records. *Theory of Probability and Its Applications*, **32**, 201–228 (English translation).

Nevzorov, V. B. and Balakrishnan, N. (1998). A record of records, In *Handbook of Statistics–16: Order Statistics: Theory and Methods*(Eds., N. Balakrishnan and C. R. Rao), pp. 515–570, North-Holland, Amsterdam, The Netherlands.

Rao, C. R. and Shanbag, D. N. (1994). *Choquet-Deny Type Functional Equations with Applications to Stochastic Models*. John Wiley & Sons, Chichester, U.K.

Wesolowski, J. and Ahsanullah, M. (1997). On characterizing distributions via linearity of regression for order statistics. *Australian Journal of Statistics*, **39**, 69–78.

Chapter 8

Generalized Pareto Distributions and Their Characterizations

MAJID ASADI

Department of Statistics,
University of Isfahan, Isfahan, Iran

CONTENTS

ABSTRACT

In reliability, extreme value theory, and other branches of applied probability and statistics, there are many situations in
which one would expect to have models involving generalized
Pareto distributions (GPDs) to provide a useful description of
the observed data. Due to the importance of this family of distributions, attempts have been made by several authors to
obtain its characteristic properties. The aim of the present
chapter is to review and extend many characterization results
on GPDs and their discrete versions in the context of reliability
and ordered random variables.

KEYWORDS AND PHRASES: Mean residual life function, characterization, truncated expectation, order statistics, record
values, Lau-Rao theorem, generalized Pareto distributions, relevation type equation, residual entropy

8.1   INTRODUCTION

Recently in reliability studies and other areas of statistics, a
more general model than that of exponential distribution has
been introduced and widely used. The model is referred to as
that of the generalized Pareto distributions (GPDs) and is defined as follows: Let X be a lifetime (non-negative) random
variable with distribution function F and survival function
$\bar{F} = 1 - F$. Then F is said to be a member of the class of GPDs
if its survival function satisfies

$$\bar{F}(x) = \left(\frac{b}{ax+b} \right)^{\frac{1}{a}+1}, \qquad x \geq 0, \qquad (8.1.1)$$

where $a > -1$ and $b > 0$. This model has been introduced by Hall
and Wellner (1981), and for $a > 0$ and $-1 < a < 0$ is,

respectively, a Pareto (Lomax) distribution and a Power distribution. Moreover, by (8.1.1) with $a = 0$, we really mean its limit as a tends to the zero; in this case, the model reduces to the exponential distribution. (Note that if $-1 < a < 0$, then the distribution is bounded above.)

In reliability theory and elsewhere, in studies of the lifetime of a component or a system, this model is frequently applied due to its appealing properties such as that it has a linear mean residual life function (MRL), the corresponding coefficient of variation of the residual life is constant, and its hazard rate is the reciprocal of a linear function. On the other hand, in the reliability studies, when robustness is required against heavier-tailed distributions, the GPDs are taken to be reasonable alternatives to an exponential distribution. This model has also been extensively used in areas such as "extreme value theory." In his paper, Dargahi-Noubari (1989) recommends the GPDs for use as the distribution of excess of observed values over an arbitrarily chosen "threshold." He points out that the GPDs arise as a limit distribution for the excess over a threshold, as the threshold increases toward the right-hand tail of the distribution. An interesting connection between the GPDs and extreme value distributions is as follows: If N has a Poisson distribution with mean λ and $\{Y_i : i = 1, 2, \ldots\}$ is a sequence of i.i.d. random variables that are independent of N and are distributed with GPD of the form (8.1.1), then

$$P(\max(Y_1, Y_2, \ldots, Y_N) \le x) = \exp\left[-\lambda \left(\frac{b}{ax+b}\right)^{\frac{1}{a}+1}\right], \quad x \ge 0.$$

If X has a GPD, then the conditional distribution of $X - x$ given $X > x$ has a GPD, when $P\{X > x\} > 0$. This property is referred to in the extreme value theory as the "threshold stability" property of GPDs. Recently, Asadi et al. (2001) have given an extended definition of the GPDs as follows:

DEFINITION 8.1.1 Suppose X is a non-negative random variable. We say that it has an adapted generalized Pareto distribution, AGPD for short, with parameter vector $(c, .)$, if there exists a constant c such that $1 + cX > 0$ almost surely and

$\frac{1}{c}\log(1+cX)$, where this latter random variable is interpreted as X when $c = 0$, is exponential or geometric.

Note that if X is continuous, then its distribution is a member of the class of GPDs.

Most of the characterization results on exponential and geometric distributions involving order statistics, record values, moments of residual life function, strong memoryless properties, or relevation type equations have natural generalizations to the AGPDs. Ferguson (1967), Nagaraja (1977), Hall and Wellner (1981), and Roy and Mukherjee (1986) are among the earlier authors characterizing GPDs, or their location variations, based on various properties. Ferguson considers (essentially) the linearity of the conditional expectation of the difference of two successive order statistics given the larger of the two order statistics; Nagaraja (1977) considers the analogous property for record values; Hall and Wellner consider the linearity of the mean residual life; and Roy and Mukherjee consider the constancy of the product of the hazard function and the mean residual life function. Ahsanullah and Wesolowski (1997) [see also Dembiska and Wesolowski (1998)] extend Ferguson's result for non-adjacent order statistics. In their monograph, Rao and Shanbhag (1994) pointed out that many characterization results of the exponential and geometric distributions are linked with the results of Choquet and Deny (1960), Deny (1961) and Lau and Rao (1982). Recently, Oakes and Dasu (1990) have obtained a characterization result of generalized Pareto distributions of the type of "lack of memory property" of the exponential distribution. Motivated by this, Asadi et al. (2001) extend and unify many characterization results of the exponential and geometric distributions, especially those that are linked with aforementioned contributions of Deny and Lau-Rao, to achieve some major results on AGPDs.

In this chapter an attempt is made to summarize many of the known results on characteristic properties of AGPD (in particular of GPD). Section 8.2 is a discussion on characterizations based on conditional expectations; we prove, in this section, a general result characterizing AGPDs based on truncated expectation. In Section 8.3 we review the characterizations based on order statistics and record values. In Section 8.4 we characterize the GPD based on the relevation type equation. In Section 8.5 we give some characterizations of the GPDs using the residual uncertainty introduced by Ebrahimi (1996).

8.2 CHARACTERIZATIONS BASED ON TRUNCATED EXPECTATIONS

Among numerous characterizations of the probability distribution of a random variable or a random vector, those based on the corresponding MRL function or, in general, on the conditional expectations

$$E\left[h(X) - h(x) \mid X \geq x\right] = g(x) \tag{8.2.1}$$

or on

$$E\left[h(X - x) \mid X \geq x\right] = E\left[h(X)\right], \tag{8.2.2}$$

with h and g meeting some conditions, have proved to be of increasing interest. Gupta (1975) proved that if h is an increasing and differentiable function, then the distribution of a non-negative continuous random variable X can be characterized by Equation (8.2.1). He also arrived at a similar result for the case when X is a discrete non-negative integer-valued random variable. Hamdan (1972) had already obtained a special case of this result of Gupta (1975). Kotz and Shanbhag (1980) generalized the Gupta (1975) result and proved that when X is an arbitrary random variable, then under some conditions, the conditional expectations $E[h(X) \mid X \geq x]$ determine the distribution of X uniquely.

Although the result of Kotz and Shanbhag (1980) shows that under some mild conditions, the conditional expectations $E[h(X) \mid X \geq x]$ determine the underlying distribution uniquely, Rao and Shanbhag (1994) gave an example showing that the situation is not the same for the conditional expectations of the form $E\left[h(X - x) \mid X \geq x\right]$. In the literature, partial characterizations based on the latter case have been considered by several authors. Shanbhag (1970) proved that for a general non-negative random variable X, Equation (8.2.2) with $h(x) = x$

is valid (i.e., the MRL of a random variable X is constant) if and only if the underlying distribution is exponential and also gave a related characterization of the geometric distribution. Azlarov et al. (1972) showed that a non-negative continuous random variable X is exponential if and only if

$$E[(X - x)^2 \mid X \geq x] = c \quad \text{for all } x \geq 0,$$

where c is a positive constant. Sahobov and Geshev (1974) extended this result to arrive at a characterization of the exponential distribution based on Equation (8.2.2) with h as a polynomial function. Dallas (1979) showed that if X is non-negative continuous and for some positive integer r, $E[X^r]$ is finite, then

$$E[(X - x)^r \mid X > x] = c \quad \text{for all } x \geq 0$$

if and only if X is exponentially distributed, where c is positive constant.

Azlarov et al. (1972) also characterized the exponential distribution using the truncated variance of a positive random variable. They showed that a positive random variable X is exponentially distributed if and only if

$$Var(X \mid X > x) = c \quad \text{for all } x \geq 0,$$

where $Var(.\mid.)$ denotes the variance of conditional distribution of X. Nagaraja (1975) proved that a positive continuous random variable X is exponential if and only if

$$\{E[X - x \mid X > x]\}^2 = Var(X \mid X > x) \quad \text{for all } x \geq 0.$$

Rao and Shanbhag (1994, 1998), using the Lau–Rao theorem, proved that a non-negative random variable X has an exponential (geometric) distribution if and only if $E[h(X - x) \mid X \geq x] = E[h(X)]$, where h satisfies some conditions. Asadi et al. (2001) have extended this result as follows:

THEOREM 8.2.1 *Let c be a real number and X be a non-negative non-degenerate continuous random variable such that $cX + 1 > 0$ almost surely. Let h be a monotonic right continuous function on R_+ such that $E(\mid h(X) \mid) < \infty$, $E(\mid h(X) \mid) \neq h(0)$, and $h(\frac{e^{cx} - 1}{c})$, $x \in R_+$ (where we take $\frac{e^{cx} - 1}{c} = x$ if $c = 0$) satisfies the condition that whenever it is arithmetic, the distribution of*

$\frac{1}{c}\log(1 + cX)$ *is also arithmetic with the same span as that of the function in question. Then*

$$E\left[h\left(\frac{X-x}{cx+1}\right)\Big| X \geq x\right] = E[h(X)], \quad x \in R_+ \quad \text{with } P\{X \geq x\} > 0$$

$$(8.2.3)$$

if and only if X is distributed with $AGPD(\frac{c}{\alpha}, .)$.

PROOF. The result for $c = 0$ is essentially the result of Rao and Shanbhag (1998). Note that, in this case, (8.2.3) is equivalent to

$$\int_{R_+} \bar{F}(y+x)\mu_{\hat{h}}(dy) = \bar{F}(x), \quad (8.2.4)$$

where $\bar{F}(x) = P\{X \geq x\}$, $x \in R_+$, and $\mu_{\hat{h}}$ is the measure determined on R_+ by $\hat{h}$ given by

$$\hat{h}(x) = \begin{cases} \frac{h(x)-h(0)}{E(h(X))-h(0)} & x \geq 0 \\ 0 & x < 0. \end{cases} \quad (8.2.5)$$

The "if" part of the result in this special case is trivial and the "only if" part follows from the Lau–Rao theorem [see Rao and Shanbhag (1994)]. In the case of $c \neq 0$, on taking

$$X^* = \frac{1}{c}\log(1 + cX),$$

$$x^* = \frac{1}{c}\log(1 + cx),$$

and

$$h^*(z) = h\left(\frac{e^{cz}-1}{c}\right), \quad z \in R_+,$$

we have (8.2.3) to be equivalent to

$$E\{h^*(X^* - x^*) \mid X^* \geq x^*\}$$
$$= E(h^*(X^*)), \quad x^* \in R_+ \quad \text{with } P\{X^* \geq x^*\} > 0. \quad (8.2.6)$$

Obviously, (8.2.6) meets the requirements of (8.2.3) for $c = 0$ with (x^*, X^*, h^*) in place of (x, X, h). Hence, we have the result. ∎

This theorem gives the following corollaries. The first one is an extension of the characterization of the exponential distribution based on a strong memoryless property.

COROLLARY 8.2.1 *Suppose X and Y are independent non-negative random variables with $P\{X \geq Y\} > P\{Y = 0\}$ and c is a real number such that $1 + cX > 0$, $1 + cY > 0$, a.s. and we have the distribution of $\frac{1}{c}\log(1 + cX)$ to be arithmetic with the same span as that of distribution of $\frac{1}{c}\log(1 + cY)$ whenever the latter is arithmetic. Then*

$$P\left\{\frac{X - Y}{cY + 1} \geq x \mid X \geq Y\right\} = P\{X \geq x\}, \quad x \in R_+$$

if and only if for some $\alpha > 0$, αX is distributed with AGPD($\frac{c}{\alpha}$, .).

PROOF. To get the result, apply Theorem 8.2.1 with $h(x) = P\{Y \leq x\}, x \in R_+$. ∎

COROLLARY 8.2.2 *A non-negative continuous random variable X has a GPD with parameters a and b if and only if for some $k > 0$*

$$E[(X - x)^k \mid X > x] = (ax + b)^k, \quad x \geq 0.$$

Note that this corollary subsumes the results of Dallas (1979) and Hall and Wellner (1981).

REMARK 8.2.1 It should be pointed out here that in the case when X has GPD, the quantity $1 + cx$ in Theorem 8.2.1 and otherwise is equal to $\frac{m(x)}{m(0)}$, where m denotes the MRL of X.

8.2.1 An extended version of the Oakes–Dasu result

Oakes and Dasu (1990) gave a characterization of the GPDs using a property of the type of "lack of memory." They showed that the survival function of the residual life distribution of an absolutely continuous random variable X, given by

$$\bar{F}(y; x) = P\{X > x + y \mid X > x\} = \frac{\bar{F}(x + y)}{\bar{F}(x)}, \quad \bar{F}(x) > 0,$$

satisfies, for some non-negative function θ on R_+, in the following

$$\bar{F}(\theta(x)y; x) = \bar{F}(y), \quad x, y \in R_+, \tag{8.2.7}$$

if and only if θ is linear and $\bar{F}$ is a GPD. Asadi et al. (2001) gave the following theorem, which proves an extended version of the result of Oakes and Dasu (1990), relaxing, in particular, the assumption of absolute continuity of $\bar{F}$:

THEOREM 8.2.2 *Let $\bar{F}$ be a survival function on R_+ (where we define it here as $\bar{F}(x) = 1 - F(x)$, with F as the distribution function concentrated on R_+) such that the corresponding distribution has a finite mean. Let $\theta : R_+ \longrightarrow R_+$. Then there exists a point $x_0 \in R_+$ with $\bar{F}(x_0) > 0$ and a sequence $\{x_n : n = 1, 2, \ldots\}$ of points lying in (x_0, ∞) such that it converges to x_0 and*

$$\bar{F}(x_n + \theta(x_n)y) = \bar{F}(x_n)\bar{F}(y), \quad n = 0, 1, \ldots; \quad y > 0 \tag{8.2.8}$$

if and only if the distribution corresponding to $\bar{F}$ is GPD(c,.) (i.e., continuous AGPD(c,.)) and $\theta(x_n) = 1 + cx_n$, $n = 0, 1, \ldots$ for some c.

8.3 CHARACTERIZATION RESULTS ON ORDER STATISTICS AND RECORD VALUES

Since the emergence of Ferguson (1967), many attempts have been made on characterizing distributions based on properties of order statistics. Galambos and Kotz (1978), David (1981), Azlarov and Volodin (1986), Arnold et al. (1992), Rao and Shanbhag (1994), and Kamps (1995), among others, have reviewed the existing literature on important characteristic properties of the order statistics and record values.

8.3.1 Characterization based on equality of distribution

Let $X_1, X_2, \ldots, X_n$ be a random sample of size n from the distribution function F. Let also $X_{1:n} \leq X_{2:n} \leq \ldots \leq X_{n:n}$ denote the corresponding order statistics. Because of the importance of the exponential distribution (and geometric distribution in the discrete case) in reliability and other areas of statistics, in the last three decades or so, researchers have obtained many characterization results on these distributions based on the order statistics. Most of these characterization results involve the spacings of the order statistics (i.e., $Z_i = X_{i+1:n} - X_{i:n}$, $i = 0, 1, \ldots, n-1$, with $X_{0:n} = 0$). Among the numerous results based on the spacings, we mention a few here.

Epstein and Sobel (1953) and Renyi (1953) showed that for an exponential distribution the sequence of the normalized spacings $D_{i,n} = (n-i)(X_{i+1:n} - X_{i:n})$, $i = 0, 1, \ldots, n-1$ is independent and identically distributed. Puri and Rubin (1970) proved that in the case when F is absolutely continuous the converse of the latter result is true for $n = 2$. Basu (1965) proved that the independence of the random variables

$$Z_1 = X_{1:n}, \quad Z_2 = X_{2:n} - X_{1:n}, \ldots, \quad Z_n = X_{n:n} - X_{n-1:n}$$

characterizes the exponential distribution. Seshadri   et al. (1969) showed that the distribution F is exponential if and only if the normalized spacings $D_{i,n}$, $1 \leq i < n$, are exponentially distributed. Tanis (1964) showed that a continuous distribution function F is exponential if and only if

$$\sum_{i=1}^{n}(X_{i:n} - X_{1:n}) \quad \text{and} \quad X_{1:n}$$

are independent. Rossberg (1972) generalized the Tanis result as follows. A continuous distribution function F is exponential if and only if for fixed $l < k$, $X_{l:k}$ and

$$W_{km} = c_k X_{k:n} + c_{k+1} X_{k+1:n} + \cdots + c_m X_{m:n}, \quad k < m \leq n$$

with $\sum_{i=k}^{m} c_i = 0$, $c_k \neq 0$, $c_m \neq 0$, are independent. Ahsanullah and Rahman (1972) showed that a continuous distribution F is exponential if and only if the order statistics $X_{k:n}$ can be expressed as

$$X_{k:n} \overset{d}{=} \sum_{i=1}^{n} \frac{Y_i}{n - i + 1}$$

where $Y_i's$ $(j = 1, 2, \ldots, k)$ are independent and identically distributed random variables with distribution function F, where d stands for the distribution.

Ahsanullah (1977) proved that under some conditions on F, we have it to be exponential if and only if the spacing $D_{n-1,n}$ is exponential for some $n \in N$. Rossberg (1972) proved that a continuous distribution function F is exponential if and only if

$$X_{i+1:n} - X_{i:n} \overset{d}{=} \min\{X_1, \ldots, X_{n-i}\}.$$

The problem of characterizations of the geometric distribution based on order statistics has also been considered in the literature. Srivastava (1974) proved that a discrete distribution F is geometric with probability function given by

$$P(X = i) = p(1-p)^{(i-\alpha)/\beta}, \quad i = \alpha, \alpha + \beta, \alpha + 2\beta, \ldots,$$

if and only if its order statistics satisfy

$$P(X_{1:n} = \alpha + i\beta, Z = 0)$$
$$= P(X_{1:n} = \alpha + i\beta)P(Z = 0), \quad i = 0, 1 \ldots,$$

where

$$Z = \sum_{i=2}^{n}(X_{i:n} - X_{1:n}).$$

Arnold and Ghosh (1976) showed that in a sample size $n = 2$, the random variable $X_{2:2} - X_{1:2}$ given $X_{2:2} > X_{1:2}$ is distributed as X if and only if X has a geometric distribution. Arnold (1980) extended this result to show that a conditional distribution of $X_{i+1:n} - X_{i:n}$ given $X_{i+1:n} > X_{i:n}$, for $1 \leq i < n$, is distributed as $X_{1:n-i}$ if and only if X has a geometric distribution. Zijlstra (1983) gave a different proof of the latter result.

Rao and Shanbhag (1994) [see also Rao and Shanbhag (1998)] pointed out that many characterization results based on order statistics and record values have implicit links with the integrated Cauchy functional equation or its variants, such as the Lau–Rao theorem. These latter authors, essentially based on the Lau–Rao theorem, proved that for some $1 \leq i < n$

$$X_{i+1:n} - X_{i:n} \overset{d}{=} \min\{X_1, \ldots, X_{n-i}\} \tag{8.3.1}$$

if and only if one of the following holds:

(i) F is exponential.
(ii) F is concentrated on some semi-lattice of the form $\{0, \lambda, 2\lambda, \ldots\}$ with $F(0) = \alpha$ and $F(j\lambda) - F((j-1)\lambda) = (1-\alpha)(1-\beta)\beta^{j-1}$ for $j = 1, 2, \ldots$ for some $\alpha \in (0, \binom{n}{i}^{-1/i}]$

and $\beta \in [0, 1)$ such that $P\{X_{i+1:n} > X_{i:n}\} = (1 - \alpha)^{n-i}$ (which holds with $\alpha = \binom{n}{i}^{-1/i}$ or $\beta = 0$ if and only if

$$F(0) - F(0-) = \binom{n}{i}^{-1/i}$$

and

$$F(\lambda) - F(\lambda-) = 1 - \binom{n}{i}^{-1/i}$$

for some $\lambda > 0$).

Characterizations of the probability distributions based on record values have also been extensively studied in the literature. Let $\{R_i; i \geq 1\}$ be a sequence of record values from a distribution function F. Tata (1969) showed that the spacings $R_{i+1} - R_i, i = 1, 2, \ldots$ form a sequence of independent random variables, and she also proved that the independence of R_1 and $R_2 - R_1$ characterizes the exponential distribution. Srivastava (1980) gave an extension of Tata's result, showing that the spacing $R_{i+1} - R_i$ is independent of R_i if and only if F is exponential. Dallas (1981) proved a generalization of Srivastava's result, giving that R_i and $R_j - R_i$, $1 \leq i < j$ (with i and j fixed), are independent if and only if F is exponential. Rao and Shanbhag (1994), using essentially the Lau–Rao theorem, extended and unified the earlier results on characteristic properties of exponential distributions based on the independence of functions of record values. These latter authors showed that for some $k \geq 1$, $R_{k+1} - R_k \overset{d}{=} X_1$ where X_1 is a random variable with distribution function F, if and only if X_1 is exponential or, for some $a > 1$, X_1 is geometric on $\{a, 2a, \ldots\}$ (i.e., $a^{-1}X_1 - 1$ is geometric in the usual sense). They have also proved that when $k_2 > k_1 \geq 1$ are fixed integers, then on some interval of type $(-\infty, a]$, with $a > 0$ the left extremity of the distribution of R_{k_1}, the conditional distribution of $R_{k_2} - R_{k_1}$ given $R_{k_1} = x$ is independent of x for almost all x if and only if F is exponential, within a shift. Asadi et al. (2001) have extended many of the above-mentioned results on characterization of exponential and geometric distributions to arrive at the new results for

AGPDs. In the process of doing so, they introduced the concept of extended neighboring order statistics as follows:

DEFINITION 8.3.1 Suppose F is a probability distribution function on the real line. Then, if X_1^* and X_2^* are real random variables (defined on a probability space) such that $X_1^* \leq X_2^*$ almost surely and

$$P\left(X_1^* \leq x_1, X_2^* > x_2\right) = \mu((-\infty, x_1])(1 - F(x_2))^{\lambda},$$

$$-\infty < x_1 \leq x_2 < \infty \qquad (8.3.2)$$

for some $\lambda > 0$ and σ−finite measure μ (on R) with $supp[\mu]$ equal to $supp[F]$ or to $[x_0, \infty) \cap supp[F]$ for some $x_0 \in R$, we call these random variables extended neighboring order statistics related to F. (In (8.3.2), we take $\infty.0 = 0$ as is usually done.)

For $n \geq 2$, if $X_{1:n}, \ldots, X_{n:n}$ are order statistics based on n independent random variables with distribution function F, we have $X_{i:n}$ and $X_{i+1:n}$ for each $i = 1, 2, \ldots, n-1$ to be such that

$$(X_{i:n}, X_{i+1:n}) \stackrel{d}{=} (X_1^*, X_2^*)$$

where X_1^* and X_2^* are as in the definition above with $\lambda = n - i$ and μ as the measure determined by $\binom{n}{i} F^i(x)$, $x \in R$. Furthermore, let F be a distribution function with its right extremity not as one of its discontinuity points. If we assume $\{R_i : i = 1, 2, \ldots\}$ is the sequence of record values relative to F, then we can observe that for any $i \geq 1$

$$(R_i, R_{i+1}) \stackrel{d}{=} (X_1^*, X_2^*)$$

where X_1^* and X_2^* meet the requirements of the definition with $\lambda = 1$ and μ as the measure satisfying

$$\mu(B) = \int_B (1 - F(x))^{-1} dP\{R_i \leq x\}$$

for each Borel set B; in this latter case, we have also $P(X_1^* = X_2^*) = 0$. The following result is given by Asadi et al. (2001).

THEOREM 8.3.1 *Let X_1^* and X_2^* be extended neighboring order statistics relative to a distribution function F such that $P(X_1^* = X_2^*) < 1$ and c be a real number such that $cX_1^* + 1$ and $cX_2^* + 1$ be positive random variables almost surely. Then the conditional* *distribution of $\frac{X_2^* - X_1^*}{cX_1^* + 1}$ given that $X_2^* - X_1^* > 0$ is, in the notation of the definition, $1 - (1 - F(x))^\lambda$, $x \in R$, if and only if either F is continuous AGPDs(c,.) or, for some $\alpha > 0$, $F(\alpha e^{c\alpha}x + \frac{e^{c\alpha}-1}{c})$, $x \in R$ is a discrete AGPD(cα, .). (Here, as well as in what follows, we define $\frac{e^{c\alpha}-1}{c} = \alpha$ if $c = 0$.)*

In view of our observations immediately after Definition 8.3.1 on order statistics and record values, the following corollaries can be obtained from Theorem 8.3.1.

COROLLARY 8.3.1 *Let $n \geq 2$ and $X_1, \ldots X_n$ be non-negative non-degenerate i.i.d. random variables with distribution function F. Let $X_{1:n}, \ldots, X_{n:n}$ be the order statistics relative to $X_i's$ and c be a real number such that $cX_i + 1 > 0$ almost surely. Then, for some $1 \leq i < n$, the conditional distribution of $\frac{X_{i+1:n} - X_{i:n}}{cX_{i+1:n} + 1}$ given that $X_{i+1:n} - X_{i:n} > 0$ is the same as the distribution of $X_{1:n-i}$, where $X_{1:n-i} = \min\{X_1, \ldots, X_{n-i}\}$, if and only if either F is a continuous AGPD(c, .) or, for some $\alpha > 0$, $F(\alpha e^{c\alpha}x + \frac{e^{c\alpha}-1}{c})$, $x \in R$, is a discrete AGPD(cα, .).*

COROLLARY 8.3.2 *Let $\{R_i : i = 1, 2, \ldots\}$ be a sequence of record values relative to a distribution function F (where the right extremity of F is not one of its discontinuity points) and c be a real value number such that $cR_i + 1 > 0$ almost surely and $cR_{i+1} + 1 > 0$ almost surely for each $i \geq 1$. Then, for some $i \geq 1$, $\frac{R_{i+1} - R_i}{cR_i + 1} \overset{d}{=} R_1$ if and only if either F is a continuous AGPD(c, .) or, for some $\alpha > 0$, $F(\alpha e^{c\alpha}x + \frac{e^{c\alpha}-1}{c})$, $x \in R$, is a discrete AGPD(cα, .).*

COROLLARY 8.3.3 *Let $n \geq 2$ and $X_1, \ldots, X_n$ be i.i.d. random variables with a distribution function F that is not concentrated on $\{0\}$. Further, let $X_{1:n}, \ldots, X_{n:n}$ be the corresponding order statistics and c be a real number such that $cX_1 + 1$ is positive almost surely. Then, for some $1 \leq i < n$,*

$$\frac{X_{i+1:n} - X_{i:n}}{cX_{i:n} + 1} \overset{d}{=} \min\{X_1, \ldots, X_{n-i}\} \qquad (8.3.3)$$

if and only if one of the following holds:

(i)   *F is a continuous AGPD(c, .).*
(ii) *F is such that $F(0) - F(0-) = \binom{n}{i}^{-1/i}$ and, for some $\lambda > 0$, $F(\lambda) - F(\lambda-) = 1 - \binom{n}{i}^{-1/i}$.*

(iii) *F is of the form*

$$F(x) = \beta F_1(x) + (1 - \beta)F_2(x), \quad x \in R,$$

where F_1 is the distribution function of the degenerate distribution at the origin, F_2 is the distribution function on R such that $F_2(\alpha e^{c\alpha}x + \frac{e^{c\alpha}-1}{c})$, $x \in R$ is a discrete AGPD$(c\alpha, .)$ for some $\alpha > 0$, and β is a real number lying in $(0, \binom{n}{i}^{\frac{-1}{i}}]$ so that $P(X_{i+1:n} > X_{i:n}) = (1 - \beta)^{n-i}$.

Corollaries 8.3.1 and 8.3.2 given above follow trivially from Theorem 8.3.1, while Corollary 8.3.3 is a slight generalization of Corollary 8.3.1 and it easily follows from Theorem 8.3.1 on noting, among other things, that (in view of the fact that F is not concentrated on $\{0\}$) (8.3.3) is equivalent to the condition that F is non-degenerate and concentrated on R_+,

$$P\left\{ \frac{X_{i+1:n} - X_{i:n}}{cX_{i:n} + 1} > y \mid X_{i+1:n} > X_{i:n} \right\}$$
$$= \left(\frac{1 - F(y)}{1 - F(0)} \right)^{n-i}, \quad y \in R_+, \tag{8.3.4}$$

and $P(X_{i+1:n} > X_{i:n}) = P(X_{1:n-i} > 0)$. (Note that in the case with $P(X_{i+1:n} > X_{i:n} > 0) > 0$, the condition, in turn, is equivalent to that with "$X_{i+1:n} > X_{i:n}$" in (8.3.4) replaced by "$X_{i+1:n} > X_{i:n} > 0$"; on the other hand, when $P(X_{i+1:n} > X_{i:n} > 0) = 0$, the condition is equivalent to that F is as in (ii) above.)

8.3.2 Characterization based on expectation of functions of order statistics

Characterizations of the exponential and geometric distributions based on conditional expectations of the functions of order statistics have also been considered in the literature. Let F be a continuous distribution function with finite mean and $X_{1:n}, X_{2:n}, \ldots, X_{n:n}$ be the corresponding ordered observations based on a random sample of size $n(\geq 2)$. Ferguson (1967) showed that if for some $1 \leq i < n$

$$E(X_{i:n} \mid X_{i+1:n} = x) = ax + b, \quad \text{for almost all } [F]x \in R,$$
$$\tag{8.3.5}$$

where $a > 0$ and b is real-valued, then F has one of the following forms to within a shift and change of scale:

(i) $F(x) = e^x$ for $x \le 0$ if $a = 1$.

(ii) $F(x) = x^\theta$ for $x \in [0, 1]$ if $a \in (0, 1)$.

(iii) $F(x) = (-x)^\theta$ for $x \le -1$ if $a > 1$,

                             where $\theta = a/[i(1 - a)]$.

Wang and Srivastava (1979) proved a generalized version of Ferguson's (1967) result to arrive at the cited distributions as follows. A continuous distribution function F with finite mean is of the form of Ferguson's (1967) result if and only if its order statistics satisfy

$$E[Z_k \mid X_{i:n} = x] = \alpha x + \beta, \quad \alpha > -1, \beta \in R, \; a.s.$$

where

$$Z_k = \frac{1}{n - k} \sum_{i=k+1}^{n} \{X_{i:n} - X_{k:n}\}, \quad k = 1, 2 \ldots, n - 1.$$

Wesolowski and Ahsanullah (1997) showed that under some conditions, Ferguson's result can be extended as follows. If

$$E(X_{i+2:n} \mid X_{i:n} = x) = ax + b, \quad a > -1, \; b > 0, \; a.s.$$

then the parent distribution is a GPD.

    Beg and Kirmani (1978) obtained, under some conditions, that a continuous distribution function F, with $F(0) = 0$ and finite second moment, is exponential if and only if

$$Var(X_{i+1:n} \mid X_{i:n} = x) = c \quad \text{for all } x > 0, \; a.s.$$

where c is a positive constant and $Var(.|.)$ denotes the conditional variance. Beg and Kirmani (1979) gave another characterization of the exponential distribution: they showed that a distribution function F is exponential if and only if its order statistics satisfy

$$E[W_{i:n(c)} \mid X_{i:n} = x] = k, \quad x \ge 0, \; a.s.$$

where

$$W_{i:n(c)} = \min\{X_{i+1:n} - X_{i:n}, c\},$$

c is a positive constant, and $k \in (0, c)$. Rao and Shanbhag (1994), via the Lau–Rao theorem, proved, under some mild conditions, that for some constant $c \neq h(0+)$

$$E[h(X_{i+1:n} - X_{i:n}) \mid X_{i:n}] = c \quad a.s. \tag{8.3.6}$$

if and only if F is an exponential.

Kirmani and Alam (1980), based on a sample of size 2, proved that a discrete distribution F is geometric if and only if

$$E(X_{2:2} \mid X_{1:2} = x) = \alpha + x, \quad x = 1, 2, \ldots, \quad a.s.$$

for some constant α. Rao and Shanbhag (1986) generalized the result of Kirmani and Alam (1980). The latter authors showed that a non-negative integer-valued random variable X has a geometric distribution if and only if

$$E[h(X_{2:2} - X_{1:2}) \mid X_{1:2} = x] = \mu, \quad x = 1, 2, \ldots, \quad a.s.$$

where h is a monotone function satisfying some conditions. Nagaraja (1988) gave the discrete analogue of Ferguson's result to arrive at a characterization of the geometric distribution. In the literature there are many characterization results of the exponential distribution based on expectations of the functions of record values. Srivastava (1978) proved that a distribution function F is exponential if and only if

$$E[R_{i+1} - R_i \mid R_i] = c, \quad a.s.$$

where c is a positive constant. Aly (1988) showed that the distribution function F is uniquely determined by the following conditional expectation,

$$E[h(R_{i+1}) \mid R_i = x] = \eta(x), \quad a.s.$$

where h and η are functions satisfying some mild conditions. Nagaraja (1977) had earlier dealt with Aly's result for $i = 1$. Gupta (1984) showed that for some real constant c,

$$E((R_{i+1} - R_i)^r \mid R_i = y) = c, \quad a.s.$$

if and only if F is exponential where i and $r, r \geq 1$ are fixed. Rao and Shanbhag (1994) proved

$$E[h(R_{k+1} - R_k) \mid R_k] = \alpha \quad a.s. \tag{8.3.7}$$

where h is a monotone function, if and only if F is exponential within a shift.

The following general theorem gives us a power tool to extend most of the characterization results on exponential distribution based on conditional expectation of functions of spacings of order statistics and record values to arrive at GPDs. The proof can be found in Asadi et al. (2001).

THEOREM 8.3.2 *Let Y and Z be independent random variables such that the support of the distribution of Y equals that of the distribution of Z, and let Y be continuous. Further, let c be a real constant such that $1 + cY > 0$, $1 + cZ > 0$ almost surely, and let h be a real monotonic function on R_+ such that $E(|h(\frac{(Y-Z)^+}{cZ+1})|) < \infty$ and $h(\frac{e^{cx}-1}{c})$, $x \in R_+$ (with $\frac{e^{cx}-1}{c}$ defined to be equal to x if $c = 0$), is non-arithmetic (or non-lattice). Then, for some constant $\alpha \neq h(0+)$,*

$$E\left[h\left(\frac{(Y-Z)^+}{cZ+1}\right) \mid Y \geq Z, Z\right] = \alpha, \quad a.s. \tag{8.3.8}$$

if and only if, for some real β such that $1 + c\beta > 0$, $\frac{1}{1+c\beta}(Y - \beta)$ has a GPD(c). (By the conditional expectation in (8.3.8), we mean the one with $I_{\{Y \geq Z\}}$ in place of $Y \geq Z$, restricted to $\{Y \geq Z\}$; the assertion of the theorem also holds if "$Y \geq Z$" is replaced by "$Y > Z$.")

REMARK 8.3.1 In Theorem 8.3.2 it is assumed that Y is continuous. A natural question to ask is what would be the conclusion result if the continuity condition were dropped. We refer the reader to Asadi et al. (2001) or Rao and Shanbhag (1998) for a discussion on this.

Now the following corollary can be obtained from Theorem 8.3.2.

COROLLARY 8.3.4 *Let X_1^* and X_2^* be extended neighboring order statistics defined in the present section with F continuous and $supp[\mu] = supp[F]$. Further, let c be a real constant such that $1 + cX_1^* > 0$, $1 + cX_2^* > 0$ almost surely, and let h be a real monotonic function on R_+ such that $E(|h(\frac{X_2^*-X_1^*}{cX_1^*+1})|) < \infty$ and $h(\frac{e^{cx}-1}{c})$, $x \in R_+$ (with $\frac{e^{cx}-1}{c}$ defined to be x if $c = 0$), is* *non-arithmetic (or non-lattice). Then, for some $\alpha \neq h(0+)$,*

$$E\left[h\left(\frac{X_2^* - X_1^*}{cX_1^* + 1}\right) \mid X_2^* > X_1^*, X_1^*\right] = \alpha \quad a.s. \tag{8.3.9}$$

if and only if, for some real β such that $1+c\beta > 0$, $F(x(1+c\beta) + \beta)$, $x \in R$, is a GPD(c,.).

We can get from Corollary (8.3.4) the following corollaries. These corollaries subsume the results of Rao and Shanbhag (1998).

COROLLARY 8.3.5 *Let $X_{1:n}, X_{2:n}, \ldots, X_{n:n}$ be order statistics from a continuous distribution function F. Let conditions of Corollary 8.3.4 hold; then, for some $\alpha \neq h(0_+)$,*

$$E\left[h\left(\frac{X_{i+1:n} - X_{i:n}}{cX_{i:n} + 1}\right) \mid X_{i+1:n} > X_{i:n}, X_{i:n}\right] = \alpha \quad a.s.$$

$$\tag{8.3.10}$$

if and only if for some β such that $1 + c\beta > 0$, $F(x(1 + c\beta) + \beta)$, $x \in R$, is a GPD(c,.).

COROLLARY 8.3.6 *Let $R_1, R_2, \ldots$ be record values from a continuous distribution function F. Let conditions of Corollary 8.3.4 hold; then, for some $\alpha \neq h(0_+)$,*

$$E\left[h\left(\frac{R_{i+1} - R_i}{cR_i + 1}\right) \mid R_{i+1} > R_i, R_i\right] = \alpha \quad a.s. \tag{8.3.11}$$

if and only if for some β such that $1 + c\beta > 0$, $F(x(1 + c\beta) + \beta)$, $x \in R$, is a GPD(c,.).

REMARK 8.3.2 In view of our observation that (in obvious notation) the generalized order statistics introduced by Kamps (1995) are certain specialized versions of the extended neighboring order statistics, we can immediately get from Theorem 8.3.1 and Corollary 8.3.4 as further corollaries the corresponding results for models such as sequential order statistics and order statistics with non-integer sample size.

Corollary 8.3.5 gives a further Corollary, extending among other things a result of Ferguson (1967); the specialized versions of these results have appeared in Rao and Shanbhag (1994).

COROLLARY 8.3.7 *Let* $X_{1:n}, X_{2:n}, \ldots, X_{n:n}, n(\geq 2),$ *be order statistics from a continuous distribution function* F. *Then, for some* $1 \leq i < n,$

$$E(X_{i+1:n} \mid X_{i:n} = x) = ax + b \quad \text{for almost all } [F]x \in R$$

$$(8.3.12)$$

with a and b as constants if and only if the distribution F *is a* $GPD(c, .)$.

REMARK 8.3.3 One could mention two further characterizations of GPD: Suppose c is a real number and F is a continuous df such that $1 + cX > 0$ a.s., where $X \sim F$, and $X_{1:n}, \ldots, X_{n:n}$ are order statistics based on a random sample of size n from F and k_1 and k_2 are fixed integers such that $n \geq k_2 > k_1 \geq 1$. Then, for a fixed $a > \infty$ the left extremity of F, we have $\frac{X_{k_2:n} - X_{k_1:n}}{cX_{k_1:n} + 1}$ and $X_{k_1:n} I_{\{X_{k_1:n} < a\}}$ to be independent if and only if for some β with $1 + c\beta > 0$, in obvious notation, $\frac{1}{1+c\beta}(X - \beta)$ has GPD $(c,.)$. An analogous result for record values (with obviously "$\infty > k_2 > k_1 \geq 1$" in place of $n \geq k_2 > k_1 \geq 1$, R_{k_1} in place of $X_{k_1:n}$ and R_{k_2} in place of $X_{k_2:n}$) also holds. These results follow from their specialized versions for $c = 0$ given in Rao and Shanbhag (1994, Chapter 8; 1998).

8.3.3 Further characterization results based on ordered random variables

In view of the arguments that we have already met in this section, it is obvious that several of the other characterizations of exponential and geometric distributions, discussed in Rao and Shanbhag (1986, 1994, 1998) and Fosam and Shanbhag (1994, 1997) and other places, could be translated into those corresponding AGPDs. The following theorems provide us with some illustrations of this. For the details of the proofs, we refer to Asadi et al. (2001).

THEOREM 8.3.3 *Let* $n \geq 2$ *and* $1 \leq k \leq n - 1$ *be integers and* $Y_1, Y_2, \ldots, Y_n$ *be independent positive random variables such that* $P\{Y_i > t\} > 0$ *for each* $t > 0$ *and* $i = 2, \ldots, n$. *Further, let* c

be a real number such that $1 + cY_i > 0$ almost surely, for each $i = 1, \ldots, n$. Then

$$P\left(\frac{Y_i - Y_{i+1}}{cY_{i+1} + 1} > y \mid Y_1 > Y_2 > \ldots > Y_n\right)$$

$$= P(Y_i > y \mid Y_1 > Y_2 > \ldots > Y_i), \quad y > 0; \; i = 1, 2, \ldots, k$$

$$(8.3.13)$$

(where the right-hand side of the identity is to be read as $P(Y_i > y)$ for $i = 1$) if and only if $Y_i, i = 1, 2, \ldots, k$, are distributed with GPD(c,.) (i.e., with continuous AGPD(c,.) possibly with different parameter vectors (c,.) for different Y_i). (The result also holds if ">" in (8.3.13) is replaced by "$\geq$".)

THEOREM 8.3.4 *Let $X_1, X_2, \ldots, X_n, n \geq 2$, be independent identically distributed positive random variables, and let $a_1, \ldots, a_n$ be positive real numbers such that $\sum_1^n a_i^{-1} = 1$. Further, let c be a real number such that $1 + cX_i > 0$ almost surely for each i. Assume that the smallest closed subgroup of R containing $\log a_i, i = 1, 2 \ldots, n$, equals R itself. Then*

$$min_{1 \leq i \leq n}\{(1 + cX_i)^{a_i/c}\} \overset{d}{=} (1 + cX_1)^{1/c}, \qquad (8.3.14)$$

where $(1 + cX_i)^{\cdot/c}$ is defined to be equal to $e^{\cdot X_i}$ if $c = 0$, if and only if X_1 is distributed according to GPD(c,.).

The following theorem is an extension of Theorem 4 of Rao and Shanbhag (1998). Note that the theorem of Rao and Shanbhag referred to follows from the specialized version of our theorem for $c = 0$ and it, in turn, extends a result of Ferguson (1964, 1965) and Crawford (1966). The Ferguson–Crawford result mentioned here is that if X and Y are independent non-degenerate random variables, then min$\{X, Y\}$ and $X - Y$ are independent if and only if, for some $\alpha > 0$ and $\beta \in R$, we have $\alpha(X - \beta)$ and $\alpha(Y - \beta)$ to be either both exponential or both geometric.

THEOREM 8.3.5 *Let c be a real number, X and Y be independent non-degenerate random variables, and y_0 be a real number such that $P\{min\{X, Y\} \leq y_0\} > 0$ and both $cX + 1$ and $cY + 1$ are positive almost surely. Let ϕ be a real-valued Borel measurable function on R such that its restriction to*

$(-\infty, y_0]$ *is non-vanishing and strictly monotonic. Then* $\frac{X-Y}{cY+1}$
and $\phi(min\{X, Y\})I_{\{min\{X,Y\}\leq y_0\}}$ *are independent if and only if
for some* $\alpha \in (0, \infty)$ *and some* $\beta \in R$ *with* $1 + c\beta > 0$, $\alpha(X - \beta)$
and $\alpha(Y - \beta)$ *have both continuous AGPDs, or both discrete
AGPDs, with parameter vector* $(\frac{c}{\alpha(1+c\beta)}, .)$ *in which case* $\frac{X-Y}{cY+1}$
and $min\{X, Y\}$ *are independent.*

COROLLARY 8.3.8 *If, in Theorem 8.3.5, X and Y are addition-
ally assumed to be identically distributed, then the assertion of
the theorem holds with*

$$\frac{|X - Y|}{min\{cX + 1, cY + 1\}}$$

in place of $\frac{X-Y}{cY+1}$.

8.4   CHARACTERIZATION OF GPDs BASED ON RELEVATION TYPE EQUATION

Let X and Y denote the lifetimes of two components with distribution functions F and G, respectively. The relevation of the
survival functions $\bar{F}(t) = 1 - F(t)$ and $\bar{G}(t) = 1 - G(t)$, which
we denote by $(\bar{F}\,\bar{G})(t)$, is given by

$$(\bar{F}\,\bar{G})(t) = \bar{F}(t) - \int_0^t \frac{\bar{G}(t)}{\bar{G}(u)}\, d\bar{F}(u), \quad t \geq 0. \tag{8.4.1}$$

The concept of relevation is introduced by Krakowski (1973). In
fact, (8.4.1) is the survival function of the time to failure of two
components where the first component on failure is replaced by
the second one of the same age. On the other hand, the
convolution of $\bar{F}$ and $\bar{G}$ is given by

$$(\bar{F} * \bar{G})(t) = \int_0^\infty \bar{G}(t - u)\, d\bar{F}(u)$$

$$= \bar{F}(t) - \int_0^t \bar{G}(t - u)\, d\bar{F}(u), \quad t \geq 0. \tag{8.4.2}$$

Note that $(\bar{F} * \bar{G})(.)$ is the survival function of the time to failure of two components where the first component on failure
is replaced by a new one. The equality of (8.4.1) and (8.4.2)
has been used for characterization of the exponential distribution by many authors. Grossward   et al. (1980) used the

equality of (8.4.1) and (8.4.2) and characterized the exponential distribution under the conditions that $\bar{G}(0) = 1$ and that
$\bar{G}(t)$ is continuous and is expressible as a power series. Westcott
(1981) used a probabilistic argument to show that Grossward
et al.'s result holds under weaker conditions. Kakosyan et al.
(1984) provided an improved version of Westcott (1981). Rao
and Shanbhag (1986), without assuming the continuity of $F(t)$
and $G(t)$, established a different improved version of the
Westcott result. Recently Lau and Prakasa Rao (1990, 1992)
obtained a result which is close to that of Kakosyan et al.
(1984). The result of Lau and Prakasa Rao (1990, 1992) is
based on the version of the Choquet–Deny functional equation.
Fosam and Shanbhag (1997) have considered the multivariate
extension of the result of Lau and Prakasa Rao (1990) to arrive
at a multivariate distribution with marginals as independent
exponential distributions.

In this section, we intend to obtain a characterization result for GPD by an equality of relevation of $\bar{F}$ and $\bar{G}$ with a
modified version of the convolution of $\bar{F}$ and $\bar{G}$.

First note that the equality of the right-hand sides of
(8.4.1) and (8.4.2) gives

$$\int_0^t \left(\frac{\bar{G}(t)}{\bar{G}(u)} - \bar{G}(t-u) \right) d\bar{F}(u) = 0, \tag{8.4.3}$$

for each t with $\bar{G}(t) > 0$. If G is assumed to be continuous and
F such that its support has 0 as a cluster point, then (8.4.3)
implies trivially that given a $t \in (0, \infty)$ such that $\bar{G}(t) > 0$,
there exists a point $u_t \in (0, t)$ such that

$$h(t) = h(u_t) + h(t - u_t)$$

(note that unless $\frac{\bar{G}(t)}{\bar{G}(u)} = \bar{G}(t-u)$ for some $u \in (0, t)$, we have
a contradiction to (8.4.3)) or, equivalently, such that

$$\frac{h(t)}{t} = \frac{u_t}{t} \cdot \frac{h(u_t)}{u_t} + \frac{(t-u_t)}{t} \cdot \frac{h(t-u_t)}{(t-u_t)}, \tag{8.4.4}$$

where $h(x) = \log \bar{G}(x)$. If we assume further that $\lim_{t \to 0^+} \frac{h(t)}{t}$
exists, then (8.4.4) yields, as seen by Lau and Prakasa Rao
(1990, 1992) or more generally by Fosam and Shanbhag (1997),
that $\frac{h(t)}{t}$ is constant, or equivalently $\bar{G}(t)$ is of the form $e^{-\lambda t}$

with $\lambda > 0$ on $(0, b)$ where b is the right extremity of G; in view of the continuity of G, we then get that G is indeed exponential, i.e., we have here $b = \infty$. As (8.4.3) holds trivially when G is exponential, we have then under the assumptions that G is continuous satisfying that $\lim_{t \to 0} \frac{\log \bar{G}(t)}{t}$ exists and zero is a cluster point of supp$[F]$, that (8.4.3) holds for each t with $\bar{G}(t) > 0$ if and only if G is exponential.

A natural question to ask is whether the result of Lau–Prakasa Rao can be extended to GPDs. In the following, we show that the answer is positive.

Let X and Y denote the lifetimes of two components with distribution functions F and G, respectively. Let $\bar{F} \, \bar{*} \, G$ be the survival function of $Z = X + (cX + 1)Y = X + Y + cXY$ where $c \in R$. Then, it can be easily shown that

$$\bar{F} \, \bar{*} \, G(x) = \bar{F}(x) - \int_0^x \bar{G}\left(\frac{x - y}{cy + 1}\right) d\bar{F}(y) \quad x \geq 0. \quad (8.4.5)$$

Note that in the case when $c = 0$, (8.4.5) is the survival function of the convolution of X and Y, i.e., (8.4.2).

In the following we show that the equalities (8.4.5) and (8.4.1) characterize the GPD.

THEOREM 8.4.1 *Let $\bar{G}$ and $\bar{F}$ be two survival functions on $[0, \infty)$ and suppose that $\lim_{x \to 0} \frac{\log \bar{G}(x)}{\frac{1}{c} \log(cx+1)}, c \in R, cx + 1 > 0$, exists. Further suppose that $\bar{G}$ is continuous and zero is a cluster point of supp(F). Then the equalities (8.4.1) and (8.4.5) imply that $\bar{G}$ is GPD(c,.).*

PROOF. Note that the equalities (8.4.1) and (8.4.5) imply

$$\int_0^x \left\{ \frac{\bar{G}(x)}{\bar{G}(y)} - \bar{G}\left(\frac{x - y}{cy + 1}\right) \right\} dF(y) = 0, \quad (8.4.6)$$

for each x with $\bar{G}(x) > 0$. Under the assumptions of the theorem, given a $t \in (0, \infty)$ such that $\bar{G}(t) > 0$, there exists a $u_t \in (0, t)$ such that we have

$$h(t) = h(u_t) + h\left(\frac{t - u_t}{cu_t + 1}\right), \quad (8.4.7)$$

where $h(x) = \log \bar{G}(x)$. Essentially, using the approach used to prove Theorem 8.2.1, on taking $t^* = \frac{1}{c} \log(ct + 1)$,

$u_t^* = \frac{1}{c}\log(cu_t + 1)$ and $h^*(t) = h(\frac{e^{ct}-1}{c})$ (where $\frac{e^{ct}-1}{c}$ is to be understood as t if $c = 0$), we can immediately see that

$$h^*(t^*) = h(t)$$

and

$$h^*(t^*) = h^*(u_t^*) + h^*(t^* - u_t^*).$$

The result now follows from the assumptions of the theorem and our observations on the proof of the case of $c = 0$. That is G is GPD and the proof is complete. ∎

8.5 CHARACTERIZATION OF GPDs BASED ON RESIDUAL UNCERTAINTY

If X is a random variable having an absolutely continuous distribution function F with probability density function f, then the entropy of the random variable X is defined as

$$H(X) = H(f) = -\int_0^\infty (\log f(x)) f(x)\, dx. \qquad (8.5.1)$$

The entropy measures the "uniformity" of a distribution. As $H(f)$ increases, $f(x)$ approaches a uniform. Consequently, the concentration of probabilities decreases and it becomes more difficult to predict an outcome of a draw from $f(x)$. In fact, a very sharply peaked distribution has a very low entropy, whereas if the probability is spread out the entropy is much higher. In this sense $H(X)$ is a measure of uncertainty associated with f.

If we think of X as the lifetime of a new unit, then $H(f)$ can be useful for measuring the associated uncertainty. However, as argued by Ebrahimi (1996), if a unit is known to have survived to age t, then $H(f)$ is no longer useful for measuring the uncertainty about the remaining lifetime of the unit. In such situations, one should instead consider

$$H(f;t) = H(X;t) = -\int_t^\infty \left(\frac{f(x)}{\bar{F}(t)}\right)\left(\log\frac{f(x)}{\bar{F}(t)}\right) dx$$

$$= 1 - \frac{1}{\bar{F}(t)}\int_t^\infty (\log \lambda_F(x)) f(x)\, dx, \qquad (8.5.2)$$

where $\bar{F}(x) = 1 - F(x)$ and $\lambda_F(x) = \frac{f(x)}{\bar{F}(x)}$ is the hazard function of X. After the unit has survived for time t, $H(f;t)$ basically measures the expected uncertainty contained in the conditional density of $X-t$ given $X > t$ about the predictability of the remaining lifetime of the unit. That is, $H(f;t)$ measures concentration of conditional probabilities. Asadi and Ebrahimi (2001) gave two characterizations of the GPD based on $H(f;t)$ as follows:

THEOREM 8.5.1 *Let X be a non-negative absolutely continuous random variable with survival function $\bar{F}(x)$, hazard rate $\lambda_F(x)$, and residual uncertainty $H(f;x)$, where f is a density function of X. Then*

$$H(f;x) = c - \log \lambda_F(x) \tag{8.5.3}$$

if and only if F is GPD with survival function of the form (8.1.1), where c is a real valued constant.

It is generally known that the mean residual life function $\delta_F(t)$, $\delta_F(t) = E(X-t \mid X > t)$, is not the same as $\frac{1}{\lambda_F(t)}$. The following theorem gives another characterization of GPD.

THEOREM 8.5.2 *Let X be a non-negative absolutely continuous random variable with survival function $\bar{F}(x)$, the mean residual life function $\delta_F(x)$, and residual uncertainty $H(f;x)$. Then*

$$H(f;x) = c + \log \delta_F(x) \tag{8.5.4}$$

if and only if F is GPD of the form (8.1.1), where c is a real valued constant.

REFERENCES

Ahsanullah, M. (1977). A characteristic property of the exponential distribution. *Annals of Statistics*, **5**, 580–582.

Ahsanullah, M. and Rahman, M. (1972). A characterization of the exponential distribution. *Journal of Applied Probability*, **9**, 457–461.

Aly, M. A. H. (1988). Some contributions to characterization theory with applications to stochastic processes. Ph.D. Thesis, University of Sheffield, Sheffield, U.K.

Arnold, B. C. (1980). Two characterizations of the geometric distribution. *Journal of Applied Probability*, **17**, 570–573.

Arnold, B. C. and Ghosh, M. (1976). A characterization of the geometric distribution by properties of order statistics. *Scandinavian Actuarial Journal*, **58**, 232–234.

Asadi, M. and Ebrahimi, N. (2000). Residual uncertainty and its characterizations in terms of hazard function and mean residual life function. *Statistics & Probability Letters*, **49**, 263–269.

Asadi, M., Rao, C. R. and Shanbhag D. N. (2001). Some unified characterization results on generalized Pareto distributions. *Journal of Statistical Planning and Inference*, **93**, 29–50.

Azlarov, T. A., Dzamirzaev, A. A. and Sultanava, M. M. (1972). Characterization properties of the exponential distribution and their stability. *Random Processes and Statistical Inference*, No. II, Izdat. "Fan" Uzbek. SSR, Tashkent, 10–19.

Azlarov, T.A. and Volodin, N.A. (1986). *Characterization Problems Associated with the Exponential Distribution*. Springer-Verlag, New York.

Basu, A. P. (1965). On the characterization of the exponential distribution by order statistics. *Annals of the Institute of Statistical Mathematics*, **17**, 93–96.

Beg, M. I. and Kirmani, S. N. U. A. (1974). Characterization of the exponential distribution by a property of truncated spacing. *Sankhyā, Series A*, **41**, 278–284.

Choquet, G. and Deny, J. (1960). Sur l'équation de convolution $\mu = \mu * \sigma$. *Comptes Rendus, Academy of Sciences*, **250**, 799–801.

Crawford, G. B. (1966). Characterization of geometric and exponential distributions. *Annals of Mathematical Statistics*, **37**, 1790–1795.

Dallas, A. C. (1979). On the exponential law. *Metrika*, **26**, 105–106.

Dallas, A. C. (1981). Record values and exponential distribution. *Journal of Applied Probability*, **18**, 949–951.

Dargahi-Noubari, G. R. (1989). On tail estimation: An improved method. *Mathematical Geology*, **21**, 829–842.

David, H. A. (1981). *Order Statistics*, Second Edition. John Wiley & Sons, New York.

Deny, J. (1961). Sur l'équation de convolution $\mu = \mu^*\sigma$. *Semin. Theory Potent. M. Brelot. Fac. Sci. Paris, 1959–1960*, 4e annee.

Ebrahimi, N. (1996). How to measure uncertainty in the residual life distributions. *Sankhyā, Series A*, **58**, 48–57.

Epstein, B. and Sobel, M. (1953). Life testing. *Journal of the American Statistical Association*, **48**, 486–502.

Ferguson, T. S. (1964). A characterization of the exponential distribution. *Annals of Mathematical Statistics*, **35**, 1199–1207.

Ferguson, T. S. (1965). A characterization of the geometric distribution. *American Mathematical Monthly*, **72**, 256–260.

Ferguson, T. S. (1967). On characterizing distributions by properties of order statistics. *Sankhyā, Series A*, **29**, 265–277.

Fosam, E. B. and Shanbhag, D. N. (1994). Certain characterizations of exponential and geometric distributions. *Journal of the Royal Statistical Society, Series B*, **56**, 157–160.

Fosam, E. B. and Shanbhag, D. N. (1997). Variants of the Choquet–Deny theorem with applications. *Journal of Applied Probability*, **34**, 101–106.

Galambos, J. and Kotz, S. (1978). Characterizations of probability distributions. *Lecture Notes in Mathematics, No. 675*, Springer-Verlag, Berlin.

Gather, U. (1989). On a characterization of the exponential distribution by properties of order statistics. *Statistics & Probability Letters*, **7**, 93–93.

Grossward, E., Kotz, S. and Johnson, N. L. (1980). Characterization of the exponential distribution by relevation-type equations. *Journal of Applied Probability*, **17**, 874–877.

Gupta, R. C. (1975). On the characterization of distributions by conditional expectations. *Communications in Statistics—Theory and Methods*, **4**, 99–103.

Gupta, R. C. (1984). Relationships between order statistics and record values and some characterization results. *Journal of Applied Probability*, **21**, 425–430.

Hall, W. J. and Wellner, J. A. (1981). Mean residual life. In *Statistics and Related Topics* (Eds., M. Csorgo, D. A. Dawson, J. N. K. Rao, and A. K. Md. E. Saleh), pp. 169–184, North Holland, Amsterdam, The Netherlands.

Hamdan, M. A. (1972). On a characterization by conditional expectation. *Technometrics*, **14**, 497–499.

Huang, J. S. (1978). On a "lack of memory" property. *Statistical Technical Report*, University of Guelph, Guelph, Ontario, Canada.

Kakosyan A. V., Klebanov L. B. and Melamed J. A. (1984). *Characterization of Distributions by Method of Intensively Monotonic Operations*. Lecture Notes in Mathematics, No. 1088, Springer-Verlag, New York.

Kamps, U. (1995). *A Concept of Generalized Order Statistics*. B. G. Teubner, Stuttgart, Germany.

Kirmani, S. N. U. A. and Alam, S. N. (1980). Characterization of the geometric distribution by the form of a predictor. *Communications in Statistics—Theory and Methods*, **9**, 541–548.

Klebanov, L. B. (1980). Some results connected with characterization of the exponential distributions. *Theor. Veoj. i. Primeneniya*, **25**, 628–633.

Kotz, S. and Shanbhag, D. N. (1980). Some new approaches to probability distributions. *Advances in Applied Probability*, **12**, 903–921.

Krakowski, M. (1973). The relevation transform and a generalization of the gamma distribution. *Franc. Austom. Inf. Rech. Operat.*, **7**, 107–120.

Lau, K. and Rao, C. R. (1982). Integrated Cauchy functional equation and characterization of exponential law. *Sankhyā, Series A*, **44**, 72–90.

Lau, K. and Prakasa Rao, B. L. S. (1990). Characterization of the exponential distribution by relevation transform. *Journal of Applied Probability*, **27**, 726–729.

Lau, K. and Prakasa Rao, B. L. S. (1992). Characterization of the exponential distribution by relevation transform. *Journal of Applied Probability*, **29**, 1003–1004.

Nagaraja, H. N. (1975). Characterization of some distributions by conditional expectations. *Journal of Indian Statistical Society*, **13**, 57–61.

Nagaraja, H. N. (1977). On a characterization based on record values, *Australian Journal of Statistics*, **16**, 70–73.

Nagaraja, H. N. (1988). Record values and related statistics—a review. *Communications in Statistics—Theory and Methods*, **17**, 2223–2238.

Oakes, D. and Dasu, T. (1990) A note on residual life. *Biometrika*, **77**, 409–410.

Puri, P. S. and Rubin, H. (1970). A characterization based on the absolute difference of two i.i.d. random variables. *Annals of Mathematical Statistics*, **41**, 251–255.

Rao, C. R. and Shanbhag, D. N. (1986). Recent results on characterization of probability distributions: A unified approach through extensions of Deny's theorem. *Advances in Applied Probability*, **18**, 660–678.

Rao, C. R. and Shanbhag, D. N. (1994). *Choquet–Deny Type Functional Equations with Applications to Stochastic Models*. John Wiley & Sons, Chichester, U.K.

Rao, C. R. and Shanbhag, D. N. (1998). Recent approaches to characterizations based on order statistics and record values. In *Handbook of Statistics, Vol. 16: Order Statistics: Theory and Methods* (Eds., N. Balakrishnan and C. R. Rao), pp. 231–256, North-Holland, Amsterdam, The Netherlands.

Renyi, A. (1953). On the theory of order statistics. *Acta. Math. Acad. Sci. Hung.*, **4**, 191–231.

Rossberg, H. J. (1972). Characterization of the exponential and Pareto distributions by means of some properties of the distributions which the difference and quotients of the order statistics are subject to. *Mathematisch Operatonsforchung und Statistik*, **3**, 207–216.

Roy, D. and Mukherjee, S. P. (1986). Some characterizations of exponential and related life distributions. *Calcutta Statistical Association Bulletin*, **35**, 189–197.

Sahobov, O. M. and Geshev, A. A. (1974). Characteristic property of the exponential distribution. *Natura. Univ. Plovdiv.*, **7**, 25–28 (in Russian).

Shanbhag, D. N. (1991). Extended versions of Deny's theorem via de Finetti's theorem. *Computational Statistics & Data Analysis*, **12**, 115–126.

Shimizu, R. (1979). On a lack of memory of the exponential distribution. *Annals of the Institute of Statistical Mathematics*, **39**, 309–313.

Smith, R. L. (1987). Estimating tails of probability distributions. *Annals of Statistics*, **15**, 1174–1207.

Srivastava, R. C. (1974). Two characterizations of the geometric distribution. *Journal of the American Statistical Association*, **69**, 267–269.

Tanis, E. A. (1964). Linear forms in the order statistics from an exponential distribution. *Annals of Mathematical Statistics*, **35**, 270–276.

Tata, M. N. (1969). On outstanding values in a sequence of random variables. *Z. Wahrsch. Verw.*, **12**, 9–20.

Wang, Y. H. and Srivastava, R. C. (1980). A characterization of the exponential and related distributions by linear regression. *Annals of Statistics*, **8**, 217–220.

Westcott M. (1981). Letter to the editor. *Journal of Applied Probability*, **18**, 568.

Wesolowski, J. and Ahsanullah, M. (1997). On characterizing distributions via linearity of regression for order statistics. *Australian Journal of Statistics*, **39**, 69–78.

Wesolowski, J. and Dembinska, A. (1999). Linearity of regression for non-adjacent statistics. *Metrika*, **48**, 215–225.

Zijlstra, M. (1983). Characterizations of the geometric distribution by distribution properties. *Journal of Applied Probability*, **20**, 843–850.

Chapter 9

On Some Characteristic Properties of the Uniform Distribution

G. ARSLAN
Department of Statistics and Computer Science, Başkent University, Ankara, Turkey

M. AHSANULLAH
Department of Management Sciences, Rider University, Lawrenceville, NJ, U.S.A.

I. G. BAIRAMOV
Department of Mathematics, İzmir University of Economics, İzmir, Turkey

CONTENTS

ABSTRACT

In this paper some characteristic properties of the uniform distribution will be considered. Let $V_1, V_2, \ldots, V_n$ be independent and identically distributed random variables with distribution function $F_{V_i}(x) = x^i$. Denote by $X_{L(n)}$ the nth lower record value of the sequence $\{X_j,\ j = 1,\ 2,\ \ldots\}$ and by $X_{1,n}$ the first-order statistics of $\{X_1,\ X_2,\ \ldots,\ X_n\}$. It will be shown that the relations

$$X_{1,n} \stackrel{d}{=} X_{1,n-1} V_n \quad \text{and} \quad X_{L(n)} \stackrel{d}{=} X_{L(n-1)} V_1$$

are characteristic properties of the uniform distribution.

KEYWORDS AND PHRASES: Order statistics, record values, characterization, uniform distribution

9.1   INTRODUCTION

Let $X_1, X_2, \ldots, X_n$ be independent and identically (iid) distributed random variables with a continuous distribution function F; then the probability density function of the ith order statistics $X_{i,n}, i = 1, 2, \ldots, n$ is given by

$$f_{i,n}(x) = \frac{n!}{(i-1)!(n-i)!} [F(x)]^{i-1}$$

$$\times [1 - F(x)]^{n-i} f(x), \quad -\infty < x < \infty. \qquad (9.1.1)$$

For various properties of order statistics see Ahsanullah and Nevzorov (2002), Arnold pe et al. (1992), and David (1981). Let $X_1, X_2, \ldots$ be a sequence of random variables. The lower record values of this sequence can be defined in the following way. Let $Y_1 = X_1$ and $Y_n = \min\{X_1, \ldots, X_n\}$ for $n > 1$. Then $X_j, j > 1$ is called a lower record value of the sequence $\{X_i, i = 1, 2, \ldots\}$ if $Y_j < Y_{j-1}$. The lower record times are defined as $L(n) = \min\{j \mid j > L(n-1), X_j < X_{L(n-1)}\}$ with $L(1) = 1$ and $n > 1$. If the random variables are iid with a continuous distribution function F, then the probability density function of the nth lower record value $X_{L(n)}$ is given by

$$f_{L(n)}(x) = \frac{[H(x)]^{n-1}}{(n-1)!} f(x), \quad -\infty < x < \infty, \qquad (9.1.2)$$

where $H(x) = -\ln F(x)$.

There are many interesting papers on characterizations involving order statistics and record values. See, for example, Ahsanullah (1977, 1995), Bairamov (2000), Bairamov and Apaydin (2000), Dembinska and Wesolowski (1998, 2000), and the references in these papers.

Let ε_i for $i = 1, \ldots, m$ and $X_1, X_2, \ldots, X_n, \ldots$ be iid uniformly distributed random variables on $(0, 1)$. Then it is known that [see Ahsanullah (1995)]

$$X_{L(m)} \overset{d}{=} \varepsilon_1 \cdots \varepsilon_m. \tag{9.1.3}$$

Similarly, if $X_1, X_2, \ldots, X_n$ are iid uniform random variables on $(0, 1)$ and $V_1, \ldots, V_n$ are independent random variables with distribution function

$$F_{V_i}(x) = x^i, \quad 0 < x < 1, \tag{9.1.4}$$

then it can be shown that

$$X_{1,n} \overset{d}{=} V_1 V_2 \cdots V_n. \tag{9.1.5}$$

Bairamov and Arslan (2000) have shown that relation (9.1.3) holds as well if it is only known that $X_1, X_2, \ldots, X_n, \ldots$ are non-negative iid random variables. In a similar way, it can be shown that if $X_1, X_2, \ldots, X_n$ are non-negative iid random variables, then relation (9.1.5) is also true.

In this paper we will present some characterizations of the uniform distribution using order statistics and record values with relations similar to (9.1.5).

9.2 RESULTS

The first result gives a characterization using lower record values, while in the second result order statistics are used. We will write $X \in U(0, 1)$ if X is a random variable uniformly distributed on $(0, 1)$.

THEOREM 9.2.1 *Let $X_1, X_2,.$ be a sequence of iid non-negative absolutely continuous bounded random variables and let V_1 be uniformly distributed on $(0, 1)$ and independent of the X_i's. Without loss of generality, assume that $F(0) = 0$ and $F(1) = 1$.*

Then the relation

$$X_{L(n)} \stackrel{d}{=} X_{L(n-1)} V_1 \tag{9.2.1}$$

holds for some fixed $n > 1$ if and only if $X_i \in U(0, 1)$.

PROOF. It is well known [see Ahsanullah (1995)] that if $X_i \in U(0, 1)$, then $X_{L(n)} \stackrel{d}{=} X_{L(n-1)} V_1$.

Suppose that $X_{L(n)} \stackrel{d}{=} X_{L(n-1)} V_1$. Then

$$F_{L(n)}(x) = P(X_{L(n)} \le x) = \int_0^1 F_{L(n-1)}(x/u)\, du$$

$$= \int_0^x du + \int_x^1 F_{L(n-1)}(x/u)\, du$$

$$= x + x \int_x^1 F_{L(n-1)}(t)\, t^{-2}\, dt. \tag{9.2.2}$$

Differentiating both sides of (9.2.2) with respect to x, we obtain

$$f_{L(n)}(x) = 1 - \frac{1}{x} F_{L(n-1)}(x) + \int_x^1 F_{L(n-1)}(t)\, t^{-2}\, dt. \tag{9.2.3}$$

Thus,

$$x f_{L(n)}(x) = x - F_{L(n-1)}(x) + x \int_x^1 F_{L(n-1)}(t)\, t^{-2}\, dt$$

$$= x - F_{L(n-1)}(x) + F_{L(n)}(x) - x$$

$$= F_{L(n)}(x) - F_{L(n-1)}(x). \tag{9.2.4}$$

Now,

$$F_{L(n)}(x) - F_{L(n-1)}(x)$$

$$= \int_0^x \frac{(H(u))^{n-1}}{\Gamma(n)} f(u)\, du - \int_0^x \frac{(H(u))^{n-2}}{\Gamma(n-1)} f(u)\, du$$

$$= F(u) \frac{(H(u))^{n-1}}{\Gamma(n)} \Big|_0^x + \int_0^x \frac{(H(u))^{n-2}}{\Gamma(n-1)} f(u)\, du$$

$$- \int_0^x \frac{(H(u))^{n-2}}{\Gamma(n-1)} f(u)\, du$$

$$= F(x) \frac{(H(x))^{n-1}}{\Gamma(n)}$$

$$= \frac{F(x)}{f(x)} f_{L(n)}(x). \tag{9.2.5}$$

Combining (9.2.4) and (9.2.5), we have

$$\frac{f(x)}{F(x)} = \frac{1}{x}, x = 0. \tag{9.2.6}$$

The solution of (9.2.6) is

$$F(x) = cx, \tag{9.2.7}$$

where c is a constant. Since $F(0) = 0$ and $F(1) = 1$, we must have

$$F(x) = x , \ 0 \le x \le 1. \tag{9.2.8}$$

∎

Before stating the second result, we note that Nevzorov (2001) gave the following relation of the uniform distribution.

Let X_i $(i = 1, 2, \ldots, n)$ and W_i $(i = 1, 2, \ldots, n)$ be independent and identically distributed uniform, $U(0, 1)$, random variables and let W_i $(i = 1, 2, \ldots, n)$ be independent of the X_i's. Then the following relation holds:

$$X_{k,n} \overset{d}{=} X_{k,m} W_{m+1,n}. \tag{9.2.9}$$

Wesolowski and Ahsanullah (2004) have obtained a general characterization result of the power function distribution based on k order statistics using this property. We present here a characterization of the uniform distribution based on the first order statistics.

THEOREM 9.2.2 *Let $X_1, X_2, \ldots, X_n$ be iid non-negative absolutely continuous bounded random variables and let V_n be independent of the X_i's with distribution function $F_{V_n}(x) = x^n$, $0 < x < 1$. Without any loss of generality, assume that $F(0) = 0$ and $F(1) = 1$. Then the condition*

$$X_{1,n} \overset{d}{=} X_{1,n-1} V_n \tag{9.2.10}$$

holds for some fixed $n > 1$ if and only if $X_i \in U(0, 1)$.

PROOF. It is easy to show that if $X_i \in U(0, 1)$, then $X_{1,n}$ is identically distributed as $X_{1,n-1} V_n$.

If (9.2.10) holds for some positive integer n, then

$$F_{1,n}(x) = \int_0^1 F_{1,n-1}(x/u)nu^{n-1}du$$

$$= n\int_0^x u^{n-1}du + n\int_x^1 F_{1,n-1}(x/u)\,u^{n-1}du$$

$$= x^n + nx^n \int_x^1 F_{1,n-1}(t)\,t^{-n-1}dt. \qquad (9.2.11)$$

Differentiating (9.2.11) with respect to x, we obtain

$$f_{1,n}(x) = nx^{n-1} - nx^{-1}F_{1,n-1}(x)$$

$$+n^2x^{n-1}\int_x^1 F_{1,n-1}(t)t^{-n-1}\,dt. \qquad (9.2.12)$$

Using (9.2.11), we obtain

$$f_{1,n}(x) = nx^{n-1} - nx^{-1}F_{1,n-1}(x)$$

$$+nx^{-1}(F_{1,n}(x) - x^n), \qquad (9.2.13)$$

where $f_{1,k}(x)$ is the pdf of $X_{1,k}$. On simplification, we have

$$xf_{1,n}(x) = n\{F_{1,n}(x) - F_{1,n-1}(x)\}. \qquad (9.2.14)$$

Now

$$n[F_{1,n}(x) - F_{1,n-1}(x)]$$

$$= n[\bar{F}_{1,n-1}(x) - \bar{F}_{1,n}(x)]$$

$$= n\left[(\bar{F}(x))^{n-1} - (\bar{F}(x))^{n-1}(1 - F(x))\right]$$

$$= n[\bar{F}(x)]^{n-1}F(x)$$

$$= \frac{F(x)}{f(x)}f_{1,n(x)}, \qquad (9.2.15)$$

where $\bar{F} = 1 - F$. Combining (9.2.14) and (9.2.15), we obtain

$$\frac{f(x)}{F(x)} = \frac{1}{x}, x = 0, \qquad (9.2.16)$$

which is the same as Equation (9.2.6). Hence, $F(x) = x, 0 \le x \le 1$. ∎

9.3   CONCLUSIONS AND REMARKS

As can be seen from the theorems obtained in this paper, investigating relations similar to

$$X_{L(m)} \overset{d}{=} \varepsilon_1 \cdots \varepsilon_m \tag{9.3.1}$$

or

$$X_{1,n} \overset{d}{=} V_1 V_2 \cdots V_n \tag{9.3.2}$$

may lead to interesting characterization results. In this respect, the authors have studied another relation involving products of random variables.

Let $\varepsilon_1, \ldots, \varepsilon_m$ be iid random variables such that

$$F_{\varepsilon_i}(x) = \begin{cases} 1 - x^{-\beta^{-1}}, & x \geq 1 & if \quad \beta > 0 \\ x^{-\beta^{-1}}, & 0 < x < 1 & if \quad \beta < 0. \end{cases} \tag{9.3.3}$$

If $X_1, X_2, \ldots, X_n, \ldots$ are non-negative iid random variables, then

$$X_{U(m)} \overset{d}{=} \frac{1}{\beta} \varepsilon_1 \cdots \varepsilon_m, \beta \neq 0, \tag{9.3.4}$$

if and only if

$$F(x) = 1 - (1 + \beta x)^{-\beta^{-1}},$$
$$\begin{cases} x \geq 0 & if \quad \beta > 0 \\ 0 > x < -1 & if \quad \beta < 0. \end{cases} \tag{9.3.5}$$

It is also interesting to note that relation (9.3.4) includes relation (9.3.1) as a special case ($\beta = -1$).

REFERENCES

Ahsanullah, M. (1977). A characteristic property of the exponential distribution. *Annals of Statistics*, **5**, 580–582.

Ahsanullah, M. (1995). *Record Statistics*. Nova Science Publishers, Commack, New York.

Ahsanullah, M. and Nevzorov, V. (2002). *Ordered Random Variables*. Nova Science Publishers, Commack, New York.

Arnold, B. C., Balakrishnan, N. and Nagaraja, H. N. (1992). *A First Course in Order Statistics*. John Wiley & Sons, New York.

Arslan, G. (2001). *Characterization of Probability Distributions*. Ph.D. Thesis, Anadolu University, Department of Statistics, Turkey.

Bairamov, I. G. (2000). On the characteristic properties of the exponential distribution. *Annals of the Institute of Statistical Mathematics*, **52**, 448–458.

Bairamov, I. G. and Apaydin, A. (2000). Characterization of continuous distributions by property of conditional expectation. *South African Statistical Journal*, **34**, 39–50.

Bairamov, I. G. and Arslan, G. (2001). Aşagı rekorlar ile iki dağılmın karakterizasyonu, II. *İstatistik Kongresi*, Belek-Antalya, Turkey.

David, H. A. (1981). *Order Statistics*, Second edition. John Wiley & Sons, New York.

Dembinska, A. and Wesolowski, J. (1998). Linearity of regression for non-adjacent order statistics. *Metrika*, **48**, 215–222.

Dembinska, A. and Wesolowski, J. (2000). Linearity of regression for non-adjacent record values. *Journal of Statistical Planning and Inference*, **90**, 195–205.

Nevzorov, V. B. (2001). *Records. Mathematical Theory*. Translation of Mathematical Monographs, **194**, American Mathematical Society, Providence, Rhode Island.

Wesolowski, J. and Ahsanullah, M. (2004). Switching order statistics through random power contractions. *Australian and New Zealand Journal of Statistics* **46**, 297–303.

Chapter 10

Characterizations of Multivariate Distributions Involving Conditional Specification and/or Hidden Truncation

BARRY C. ARNOLD

*Department of Statistics, University of California,
Riverside, CA, U.S.A.*

CONTENTS

ABSTRACT

Conditionally specified distributions are ones which have the property that their conditional densities belong to specified parametric families. Hidden truncation models introduce skewness into classical models by conditioning on events involving an unobserved threshhold variable. Both kinds of models include classical models as special cases. We discuss the nature of additional assumptions needed to characterize classical models among their more flexible counterpart models.

KEYWORDS AND PHRASES: Conditionally specified models, hidden truncation, skewness, skewed normal distribution, multivariate normal distribution, Cauchy distribution, Pareto distribution

10.1 INTRODUCTION

The catalogue of available parametric families of distributions for modeling multivariate data is expanding considerably. Classical multivariate models include ones with normal, exponential, Pareto, and other marginals. Some, but not all, of these distributions have conditional distributions that are members of well-known parametric families. The classical models are often special cases of augmented families developed using a variety of techniques. The present paper focuses on two model-generating mechanisms (to be described in more detail below): conditional specification and hidden truncation. Conditional specification adds to classical models an increment of flexibility with regard to the acceptable conditional distributions in the model. Hidden truncation permits introduction of additional parameters which control skewness (or, if you wish, which introduce asymmetries into the models). Typically, the classical models are simpler but involve more distributional assumptions than do the more flexible models. We discuss the nature of the additional assumptions needed to characterize the classical models among their more flexible counterparts. Particular attention will be paid to the classical normal model and its related extensions. The nature of the assumptions necessary to characterize a distribution will, of course, be of great interest to any researcher seeking to use that distribution to model some real-world data configuration. To avoid illogical assumptions and, perhaps potentially more embarrassingly, to avoid impossible models it is necessary to ask whether the required distributional assumptions are indeed plausibly true for the data set at hand. We begin with a brief review of the classical multivariate normal distribution. Its curious restrictive properties will be used to suggest related, more flexible models for modeling multivariate data.

10.2 THE CURIOUS CLASSICAL MULTIVARIATE NORMAL MODEL

A k-dimensional random vector $\mathbf{X} = (X_1, \ldots, X_k)$ is said to have a classical multivariate normal distribution if it admits the representation:

$$\mathbf{X} = \mu + \Sigma^{1/2}\mathbf{Z} \tag{10.2.1}$$

where $\mu \in \mathbf{R}^k, \Sigma$ is positive definite, and the coordinates of $\mathbf{Z} = (Z_1, \ldots, Z_k)$ are independent identically distributed standard normal random variables. In such a case, we write $\mathbf{X} \sim N^{(k)}(\mu, \Sigma)$. A random vector $\mathbf{X}$ constructed in this fashion has remarkable properties. A partial list will include the following:

1. All one-dimensional marginal distributions are normal.
2. All k_1-dimensional marginal distributions $(k_1 < k)$ are themselves k_1-variate normal distributions.
3. All linear combinations of the X_i's are normally distributed and, more generally, for an $\ell \times k$ matrix $B(\ell \leq k)$ we have

$$BX \sim N^{(\ell)}(B\mu, B\Sigma B').$$

4. All conditional distributions are normal. Thus, if we partition $\mathbf{X}$ as $(\dot{\mathbf{X}}, \ddot{\mathbf{X}})'$ where $\dot{\mathbf{X}}$ is of dimension $\dot{k}$, then for every $\ddot{\mathbf{X}} \in \mathbf{R}^{k-\dot{k}}$

$$\dot{\mathbf{X}} \mid \ddot{\mathbf{X}} = \ddot{\mathbf{x}} \sim N^{(\dot{k})}(\tilde{\mu}(\ddot{\mathbf{x}}), \tilde{\Sigma}(\ddot{\mathbf{x}})).$$

5. The regression functions are linear; i.e., the coordinates of $\tilde{\mu}(\ddot{\mathbf{x}})$ in condition 4 are linear functions of $\ddot{\mathbf{x}}$.
6. Conditional variances and covariances are constants; i.e., the elements of $\tilde{\Sigma}(\ddot{\mathbf{x}})$ in condition 4 do not depend on $\ddot{\mathbf{x}}$.
7. The density of $\mathbf{X}$ is elliptically contoured; i.e., $\mathbf{X} = \mathbf{b} + \mathbf{A}\mathbf{Z}$ where $\mathbf{Z}$ is spherically symmetric.
8. $\mathbf{X}$ has linear structure; i.e., $\mathbf{X} = \mathbf{b} + \mathbf{A}\mathbf{Z}$ where the Z_i's are independent.

Individually, these properties will usually fail to characterize the classical normal model. In fact, only conditions 3 and 4 will do so (3 does so trivially, since it includes the assumption that $\mathbf{X}$ is k-variate normal). The list of conditions is important because it behooves the investigator to ask whether he believes all eight conditions are appropriate for his data configuration before he adopts a classical multivariate model. It may be remarked that condition 3, "all linear combinations of the X_i's are normal", is a particularly strong assumption and may not be at all obviously true in any particular situation. (A cynic, after viewing this property, might conclude that it is unlikely that we will encounter any classical multivariate normal data in the real world!)

In Section 10.3, we will consider the implications of variations of condition 4 (normal conditionals), what they imply, and what additional assumptions, if any, are needed to characterize the classical normal model.

10.3 NORMAL CONDITIONAL DISTRIBUTIONS

In addition to the $(\dot{\mathbf{X}}, \ddot{\mathbf{X}})'$ notation for partitioning $\mathbf{X}$ introduced earlier, we will also need:

$$\mathbf{X}_{(i)} = \mathbf{X} \text{ with } X_i \text{ deleted}, \quad i = 1, 2, \ldots, k$$

and

$$\mathbf{X}_{(i,j)} = \mathbf{X} \text{ with } X_i \text{ and } X_j \text{ deleted}.$$

Bhattacharyya (1943) discussed distributions for **X** such that for every i and every $\mathbf{x}_{(i)} \in \mathbf{R}^{k-1}$ the conditional distribution of X_i given $\mathbf{X}_{(i)} = \mathbf{x}_{(i)}$ is normal with parameters which may depend on $\mathbf{x}_{(i)}$. Arguments involving a well-known functional equation [see Arnold, Castillo and Sarabia (1999) for details] may be used to conclude that the density of such a random vector with univariate normal conditionals is of the form

$$f_{\mathbf{X}}(\mathbf{x}) = \exp\left[-\frac{1}{2}G(\mathbf{x})\right] \tag{10.3.1}$$

where

$$G(\mathbf{x}) \sum_{i_1=0}^{2} \sum_{i_2=0}^{2} \cdots \sum_{i_k=0}^{2} \gamma_{i_1 i_2,\ldots,i_k} \left[\prod_{j=1}^{k} x_j^{i_j}\right]. \tag{10.3.2}$$

This family of densities includes many densities which are not of the classical normal form. The classical multivariate normal density is associated with a choice of $G(\mathbf{x})$ in (10.3.2) that is a quadratic form in $\mathbf{x}$. Thus all of the coefficients $\gamma_{i_1 i_2,\ldots,i_k}$ for which $\sum_{j=1}^{k} i_j > 2$ must be zero. There are a variety of additional assumptions that suffice to characterize the classical multivariate model within the class of Bhattacharyya normal conditionals models [i.e., models defined by (10.3.1) and (10.3.2)]. For example, it is enough to insist that $X_i \mid \mathbf{X}_{(i)} = \mathbf{x}_{(i)} \sim N(\mu(\mathbf{x}_{(i)})\tau_i^2)$, i.e., to insist on constant conditional variances. This guarantees that $G(\mathbf{x})$ is quadratic [essentially this may be found in Bhattacharyya's (1943) paper]. Another way to guarantee that the unwanted γ's in (10.3.2) are all zero was suggested by Arnold, Castillo and Sarabia (1994). They assume that the conditional distribution of (X_i, X_j) given $\mathbf{X}_{(i,j)} = \mathbf{x}_{(i,j)}$ is classical bivariate normal for every $\mathbf{x}_{(i,j)} \in \mathbf{R}^{k-2}$, for every i, j. This guarantees that $X_i|\mathbf{X}_{(i)} = \mathbf{x}_{(i)}$ is univariate normal for each $\mathbf{x}_{(i)}$ and each i, so that (10.3.1)–(10.3.2) will hold, but in addition the unwanted γ's are also forced to be zero by the assumption that bivariate conditionals are classical bivariate normal. A related result involving only univariate conditional distributions takes the following form.

If for every i and for every $\mathbf{x}_{(i)} \in \mathbf{R}^{k-1}$,

$$X_i \mid \mathbf{X}_{(i)} = \mathbf{x}_{(i)} \sim \text{Normal}\left(\mu_i(\mathbf{x}_{(i)}), \sigma_i^2(\mathbf{x}_{(i)})\right)$$

and for every i, j and every $\mathbf{x}_{(i,j)} \in \mathbf{R}^{k-2}$

$$X_i \,|\, \mathbf{X}_{(i,j)} = \mathbf{x}_{(i,j)} \sim \text{Normal} \left(\mu_{ij}(\mathbf{x}_{(i,j)}), \sigma_{ij}(\mathbf{x}_{(ij)}) \right)$$

then $\mathbf{X}$ has a classical k-variate normal distribution. The second hypothesized condition forces the unwanted γ's to be zero here.

These results which involve restrictions on the conditional distributions can be viewed as corollaries of a marginal specification characterization.

If a k-dimensional vector $\mathbf{X}$ has a joint density of the form (10.3.1)–(10.3.2), we will write $\mathbf{X} \sim NC^{(k)}$, i.e. $\mathbf{X}$ has a k-dimensional normal conditionals distribution.

THEOREM 10.3.1 *If $\mathbf{X} \sim NC^{(k)}$ and for every i $\mathbf{X}_{(i)} \sim NC^{(k-1)}$ then $\mathbf{X}$ has a classical k-variate normal distribution (written $\mathbf{X} \sim N^{(k)}$).*

The proof of this result relies on the fact that to have the $(k-1)$ dimensional marginals of the $NC^{(k-1)}$ form, the unwanted γ's in (10.3.2) must all be zeros.

Properties 5 and 6 of the classical multivariate normal model appear to be quite restrictive. Specifically they say that for any $\ddot{\mathbf{x}}$ the conditional expectation of $\dot{\mathbf{X}}$ given $\ddot{\mathbf{X}} = \ddot{\mathbf{x}}$ is a linear function of $\ddot{\mathbf{x}}$ (i.e., we have linear regression functions) and they give the very curious condition that the conditional variance of $\dot{\mathbf{X}}$ given $\ddot{\mathbf{X}} = \ddot{\mathbf{x}}$ is a constant (i.e., does not depend on $\ddot{\mathbf{x}}$). Following Wesolowski (1991) we will say that a random vector with these two properties has Gaussian conditional structure. Do there exist random vectors which do not have a classical multivariate normal distribution but which do have Gaussian conditional structure? The answer is yes. However, it is a nontrivial exercise to characterize all such distributions. Even in two dimensions, Gaussian conditional structure can be encountered in perhaps surprising settings.

The first such bivariate example exhibiting Gaussian conditional structure (i.e., linear regressions with constant conditional variances) is due to Kwapien [see Bryc and Plucinska (1985)]. In this example the joint probability density function of the discrete random vector (X, Y) is given by

$f_{X,Y}(x, y)$:	y x	-1	-1
	-1	$\frac{1-p}{2}$	$\frac{p}{2}$
	1	$\frac{p}{2}$	$\frac{1-p}{2}$

where $p \in (0, 1)$.

Linear regression functions are automatic here, and constant conditional variances follow since $p(1 - p) = (1 - p)p$.

A higher dimensional version of the Kwapien example awaits discovery. However, Nguyen, Rempala and Wesolowski (1996) provided a simple description of k-dimensional absolutely continuous densities with Gaussian conditional structure which are not classical multivariate normal. They begin by taking $f_0(\mathbf{x})$ to be a k-dimensional classical normal density with mean vector μ_0 and variance covariance matrix Σ_0. Next, pick g_1 and g_2 to be two different densities on the interval $(0, 1)$ each with mean 0 and variance 1. Consider the joint density

$$f^*(\mathbf{x}) = f_0(\mathbf{x}) + c \prod_{i=1}^{k}(g_1(x_i) - g_2(x_i)) \tag{10.3.3}$$

where c is chosen small enough to ensure that $f^*(\mathbf{x}) > 0$, $\forall \mathbf{x}$. Evidently f^* is not a classical normal density but all of its marginals are Gaussian and all of its first and second conditional moments match those of f_0. So f^* does indeed have Gaussian conditional structure.

It is possible to characterize classical k-variate normal densities among those with Gaussian conditional structure by imposing additional conditions: for example, if we add a requirement of infinite divisibility or a requirement of elliptical symmetry. See Arnold and Wesolowski (1996) for details on this.

10.4   NON-NORMAL CONDITIONALS

Arnold, Castillo and Sarabia (1999) provide a catalog of conditionally specified models analogous to the Bhattacharyya family (10.3.1)–(10.3.2). They begin with k specific parametric families of densities on the real line. They then seek to identify all joint densities for $\mathbf{X}$ with, for each $j \neq 1, 2, \ldots, k$, all the conditional densities of X_j given $\mathbf{X}_{(j)} = \mathbf{x}_{(j)}$ being members of the jth parametric family of densities. If each of the parametric families is an exponential family of densities, then the class of all joint densities with the specific conditionals is itself an exponential family of densities. If the k specified families of univariate densities are not exponential families, it is sometimes quite difficult to determine whether any joint densities exist with the given conditional structure. However, some examples are tractable. To illustrate the nature of the results that are obtainable, we will describe two such conditionally specified families.

Suppose that we insist that for each j and for each $\mathbf{x}_j \in \mathbf{R}^{k-1}$ the conditional density of X_j given $\mathbf{X}_{(j)} = \mathbf{x}_{(j)}$ be a Cauchy distribution with location parameter $\mu_j(\mathbf{x}_{(j)})$ and scale parameter $\sigma_j(\mathbf{x}_{(j)})$, for some functions μ_j and σ_j. It may be verified that the joint density of $\mathbf{X}$ must then be of the form:

$$f_{\mathbf{X}}(\mathbf{x}) = \left[\sum_{\mathbf{i} \in T_k} m_{\mathbf{i}} \prod_{j=1}^{k} x_j^{i_j} \right]^{-1} \tag{10.4.1}$$

where T_K is the set of all vectors of 0's, 1's and 2's of dimension k.

An analogous joint density with all conditionals of the Pareto (α, σ) form can be readily constructed. Recall that X has a Pareto (α, σ) density if

$$f_X(x) = \frac{\alpha}{\sigma}(1 + \frac{x}{\sigma})^{-(\alpha+1)} \ I(x > 0) \tag{10.4.2}$$

(a convenient minor variant of the classical Pareto model). The only joint densities with all conditionals (of X_j given $X_{(j)}$ $j = 1, 2, \ldots, k$) in the family (10.4.2) are of the form:

$$f_{\mathbf{X}}(\mathbf{x}) = \left[\sum_{\mathbf{s} \in S_k} \delta_{\mathbf{s}} \left(\prod_{i=1}^{k} x_i^{s_i} \right) \right]^{-(\alpha+1)} I(\mathbf{x} > \mathbf{0}), \tag{10.4.3}$$

where S_k is the set of all vectors of 0's and 1's of dimension k.

The "classical" k-variate Cauchy distribution is included as a special case of (10.4.1). Its density takes the form

$$f_{\mathbf{X}}(\mathbf{x}) = [Q(\mathbf{x})]^{-1} \tag{10.4.4}$$

where $Q(\mathbf{x})$ is a quadratic form in $\mathbf{x}$. This model corresponds to the subclass of (10.4.1) in which $m_{\mathbf{i}} = 0$ for every $\mathbf{i}$ with $\sum_{j=1}^{k} i_j > 2$. There is an intimate relation between the normal conditionals model (10.3.1)–(10.3.2) and the Cauchy conditionals model (10.4.3). The latter can be viewed as a scale mixture of the former. Analogously, the classical k-variate Cauchy density (10.4.4) is a scale mixture of classical k-variate normal densities.

Parallel to the discussion in Section 10.3, it is possible to characterize the classical k-variate Cauchy density (10.4.2) and the classical Mardia (1962) multivariate Pareto model within the models (10.4.1) and (10.4.2) by adding additional distributional assumptions. For example, if X_i given $\mathbf{X}_{(i)} = \mathbf{x}_{(i)}$ has a univariate Cauchy distribution for every i and every $\mathbf{x}_{(i)}$ and if, in addition, (X_i, X_j) given $\mathbf{X}_{(i,j)} = \mathbf{x}_{(i,j)}$ has a bivariate classical Cauchy distribution (i.e., of the form (10.4.4) with $k = 2$), for every i, j and every $\mathbf{x}_{(i,j)}$, then the joint density of $\mathbf{X}$ must be of the classical k-variate Cauchy form [i.e., of the form (10.4.4)]. The parallel results for the Mardia k-variate Pareto distribution are discussed in Arnold, Castillo, and Sarabia (1993).

REMARK 10.4.1 The models (10.3.1)–(10.3.2), (10.4.1), and (10.4.3), are constructed to have conditional distributions in specific families. It is generally not true that their marginal densities also have conditionals in the specified families. This is true for the classical submodels but not for the general conditionally specified models. Indeed, this is another avenue to follow in characterizing the classical models.

10.5   HIDDEN TRUNCATION MODELS

The conditional specification route led us to more flexible families of densities than are available in classical models. Another fruitful strategy to invoke in the construction of flexible multivariate models involves what we may call hidden truncation. Here, too, we will begin by considering a basic normal model, recognizing that extensive generalization to other distributional bases will be feasible.

Let φ and Φ denote, respectively, the standard normal density function and distribution function. Azzalini (1985) notes that for any $\lambda \in \mathbf{R}$ the following function is a well-defined density:

$$f(x; \lambda) = 2\varphi(x)\Phi(\lambda x), \quad x \in \mathbf{R} . \tag{10.5.1}$$

Typically (when $\lambda \neq 0$) such densities will be asymmetric (skewed). Azzalini calls (10.5.1) the skew-normal density and if X has such a density we will write

$$X \sim SN(\lambda) . \tag{10.5.2}$$

Densities such as this can arise by truncation on unobserved variables (hidden truncation). For example, if in a population we believe that height and weight have a classical bivariate normal distribution, then the weights of individuals who are above average in height will have a location and scale changed version of (10.5.1) as their density function. Generalizing this model to truncation at arbitrary percentiles (instead of the median) of the unobserved variable leads to a two parameter extension of (10.5.1) that we will also call skew-normal. The density takes the form

$$f(x; \lambda_0, \lambda_1) = \varphi(x)\Phi(\lambda_0 + \lambda_1 x)/\Phi\left(\frac{\lambda_0}{\sqrt{1+\lambda_1^2}}\right) . \tag{10.5.3}$$

If X has density (10.5.2) we write

$$X \sim SN(\lambda_0, \lambda_1) .$$

A k-variate extension of this model was discussed by Arnold and Beaver (2000a). For it we have

$$f(\mathbf{x}; \lambda_0, \boldsymbol{\lambda}_1) = \left[\prod_{i=1}^{k} \varphi(x_i)\right] \Phi(\lambda_0 + \boldsymbol{\lambda}_1'\mathbf{x})\bigg/\Phi\left(\frac{\lambda_0}{\sqrt{1+\boldsymbol{\lambda}_1'\boldsymbol{\lambda}_1}}\right).$$

$$\tag{10.5.4}$$

Introduction of location and scale parameters via the formula

$$\mathbf{Y} = \mu + \Sigma^{1/2}\mathbf{X},\tag{10.5.5}$$

where $\mathbf{X}$ has density (10.5.4), $\mu \in \mathbf{R}^k$, and Σ is a $k \times k$ positive definite matrix, leads to the full k-variate skew normal model and we write

$$\mathbf{Y} \sim SN^{(k)}(\mu, \Sigma, \lambda_0, \boldsymbol{\lambda}) \ .\tag{10.5.6}$$

The study of distributional properties of this model is aided by the fact that such a random variable $\mathbf{Y}$ has a tractable moment generating function [see Arnold and Beaver (2000a) for details]. Of course, the classical k-variate normal density is included as a special case in (10.5.6). For it we merely set $\lambda_1 = \mathbf{0}$. The hidden truncation genesis of the model (10.5.5) may be reaffirmed by considering the following plausible scenario. Consider a random vector $(V_0, V_1, \ldots, V_k)$ which has a joint density of the classical $(k + 1)$-variate normal form. Now suppose that for some v_0 only observations with $V_0 < v_0$ are observed (hidden truncation on $(V_1, \ldots, V_k)$). Then the conditioned distribution of $(V_1, \ldots, V_k)$ given $V_0 < v_0$ is of the form (10.5.5).

This hidden truncation genesis for the model (10.5.5) makes transparent the following remarkable properties of the k-variate skew-normal distribution. [Detailed derivations may be found in Arnold and Beaver (2000a).]

THEOREM 10.5.1 *If* $\mathbf{X} \sim SN^{(k)}$ *[i.e., is of the form (10.5.5)–(10.5.6)], then:*

(i) *For any* $j < k$, *if* $\mathbf{X}^*$ *is a* j-*dimensional sub-vector of* $\mathbf{X}$, *then* $\mathbf{X}^* \sim SN^{(j)}$.

(ii) *If we partition* $\mathbf{X}$ *as* $(\dot{\mathbf{X}}, \ddot{\mathbf{X}})$ *where* $\dot{\mathbf{X}}$ *is of dimension* $\dot{k}$ *and* $\ddot{\mathbf{X}}$ *of dimension* $k - \dot{k}$, *then the conditional density of* $\dot{\mathbf{X}}$ *given* $\ddot{\mathbf{X}} = \ddot{\mathbf{x}}$ *is of the* $SN^{(\dot{k})}$ *form.*

In brief, all marginals **and** all conditionals of multivariate skew-normal distributions are again multivariate skew-normal. This is a sharp contrast to the Bhattacharyya normal conditionals distribution (10.3.1)–(10.3.2) which had all conditionals of the same form as the joint density but did **not** have all marginal densities of the same form (except in the special classical normal sub-case). The development of suitable techniques for estimation and inference for the k-variate skew-normal model is only in the early stages. Some discussion of estimation in the bivariate case may be found in Arnold and Beaver (2000a) and Azzalini and Capitanio (1999).

10.6   NON-NORMAL VARIANTS

Inspection of the density (10.5.4) suggests immediate generalizations. The functions φ and Φ in the numerator could be replaced by any non-normal density and any non-normal distribution. It would then only remain to determine an appropriate normalization to arrive at a related non-normal variant distribution. We can actually go further in our quest for generality, still using the hidden truncation theme as a guide.

Begin with $(k + 1)$ independent random variables V_1, $V_2, \ldots, V_k$ and U with corresponding densities and distributions given by:

$$\psi_1, \psi_2, \ldots, \psi_k, \psi_0$$

and

$$\Psi_1, \Psi_2, \ldots, \Psi_k, \Psi_0 \ .$$

Consider the conditional density of $\mathbf{V}$ given that $\lambda_0 + \lambda_1' \mathbf{V} > U$. It is readily determined that this is of the form:

$$f(\mathbf{v}; \lambda_0, \lambda_1) \propto \left[\prod_{i=1}^{k} \psi_i(v_i) \right] \Psi_0(\lambda_0 + \lambda_1' \mathbf{v}). \tag{10.6.1}$$

To make this integrate to 1, we must divide by $P(\lambda_0 + \lambda_1' \mathbf{V} > U)$. Evaluation of this quantity is easy in some cases while in other settings it will remain an awkward normalizing constant that must be numerically evaluated. It is easy to evaluate if all the V_i's and U are normal (as we have seen earlier). It is also easy if all V_i's and U are stable with the same index α (so that $\lambda_0 + \lambda_1' \mathbf{V} - U$ has a known distribution). For example, they all could be Cauchy variables. It is particularly easy to evaluate $P(\lambda_0 + \lambda_1' \mathbf{V} > U)$ if $\lambda_0 = 0$ and if all densities of $V_1, \ldots, V_k$ and U are symmetric (though possibly all different). In this case $P(\lambda'\mathbf{V} > U) = 1/2$. Some discussion of the corresponding skew-Cauchy density may be found in Arnold and Beaver (2000b).

In the development of the model (10.6.1) we conditioned on the event $\lambda_0 + \lambda_1'\mathbf{V} > U$. Instead we could condition on the complementary event $\lambda_0 + \lambda_1'\mathbf{V} < U$. The resulting joint density is only slightly changed. It is given by

$$f^*(\mathbf{v}; \lambda_0, \lambda_1) = \frac{\left[\prod_{i=1}^{k} \psi_i(v_i)\right]\left[1 - \Psi_0(\lambda_0 + \lambda_1'\mathbf{v})\right]}{P(\lambda_0 + \lambda_1'\mathbf{V} < U)}. \quad (10.6.2)$$

If Ψ_0 is symmetric this does not lead to new models not already subsumed by (10.6.1). Such would be the case in the normal and Cauchy cases. If Ψ_0 is asymmetric, then (10.6.1) and (10.6.2) define different classes of densities. Such will be the case if we focus on random variables $\mathbf{V}$ and U which are constrained to be positive. To this we turn in the next section.

10.7 SURVIVAL MODELS INVOLVING HIDDEN TRUNCATION

Consider $k + 1$ independent positive random variables V_1, $V_2, \ldots, V_k, U$ with corresponding densities and distributions denoted by

$$\psi_1, \psi_2, \ldots, \psi_k, \psi_0$$

and

$$\Psi_1, \Psi_2, \ldots, \Psi_k, \Psi_0.$$

In this setting, the conditional density of $\mathbf{V}$ given $\lambda'\mathbf{V} < U$ is of the form:

$$f(\mathbf{v}) = \frac{\left[\prod_{i=1}^{k} \psi_i(v_i)\right]\left[1 - \Psi_0(\lambda'\mathbf{v})\right]}{P(\lambda'\mathbf{V} < U)}. \quad (10.7.1)$$

Models of the form (10.7.1) will be called hidden truncation survival models. Consider the special case in which $V_i \sim$ exponential (δ_i), $i = 1, 2, \ldots, k$ and $U \sim$ exponential (δ_0). Our hidden truncation density is then given by

$$f(\mathbf{v}) \propto \left(\prod_{i=1}^{k} \delta_i e^{-\delta_i v_i} \right) e^{-\delta_0 \lambda' \mathbf{v}} \, I(\mathbf{v} > \mathbf{0}). \tag{10.7.2}$$

The density (10.7.2) factors into k functions of $v_1, v_2, \ldots, v_k$ respectively, and we see that in fact it has independent exponential $(\delta_i + \delta_0 \lambda_i)$ marginals. To get independent marginals, it did not matter what the densities $\psi_1, \psi_2, \ldots, \psi_k$ looked like; all that mattered is that U had an exponential (δ_0) density. In fact it is not difficult to verify that the hidden truncation model (10.7.1) will have independent marginals if and only if Ψ_0 is an exponential distribution function.

In some special cases, the hidden truncation survival model (10.7.1) can reduce to a joint density that just involves scale changes on the original joint density of $(V_1, \ldots, V_k)$ (with its independent marginals). Such would be the case if the V_i's had independent gamma distributions and U had an exponential distribution. In general, however, the model (10.7.1) is more flexible than the original joint density of $\mathbf{v}$.

10.8 EVEN MORE FLEXIBILITY

In the models (10.6.1) and (10.6.2) [which subsumes (10.7.1)] we could begin with an arbitrary joint density for $\mathbf{v}$ instead of one with independent marginals. Equation (10.6.2) would be replaced by

$$f(\mathbf{v}; \lambda_0, \lambda_1) = \frac{\psi(\mathbf{v}) \Psi_0 \left(\lambda_0 + \lambda_1' \mathbf{v} \right)}{P \left(\lambda_0 + \lambda_1' \mathbf{V} < U \right)}, \tag{10.8.1}$$

where $\psi(\mathbf{v})$ denotes the joint density of $(V_1, \ldots, V_k)$ which is assumed to be independent of U with distribution function Ψ_0. If the joint distribution of $\mathbf{V}$ admits the representation $\mathbf{V} = \mathbf{A} + B\tilde{\mathbf{V}}$ where $\tilde{\mathbf{V}}$ has independent marginals, then the model (10.8.1) will not lead to a genuine extension of the models obtained by location and scale transformations of variables with distributions given by (10.6.1). Otherwise, (10.8.1) will lead to new distributions. Complete generality in the hidden truncation theme would be associated with a situation in which

$(\mathbf{V}, U)$ has an arbitrary $(k + 1)$ dimensional distribution and again we condition on $\{\lambda_0 + \lambda_1' \mathbf{V} > U\}$ or $\{\lambda_0 + \lambda_1' \mathbf{V} < U\}$.

10.9   COMBINING HIDDEN TRUNCATION AND CONDITIONAL SPECIFICATION

We close by mentioning the possibility of building models which will combine the concepts of hidden truncation and conditional specification. We might consider joint densities whose conditionals belong to some hidden truncation family of densities [see, e.g., Arnold, Castillo and Sarabia (2002) for the case of skew-normal conditionals]. Alternatively we could apply hidden truncation to some conditionally specified joint density [as in (10.8.1)]. It doesn't have to stop there (though it probably should). What about a joint density obtained by applying hidden truncation to a joint density with skew-normal conditionals? Eschewing such excesses of modeling fervor, it is reasonable to argue that both hidden truncation and conditional specification provide valuable augmentations to our stock of multivariate densities to be used in modeling data configurations.

REFERENCES

Arnold, B. C. and Beaver, R. J. (2000a). Hidden truncation models. *Sankhyā, Series A*, **62**, 23–35.

Arnold, B. C. and Beaver, R. J. (2000b). The skew-Cauchy distribution. *Statistics & Probability Letters*, **49**, 285–290.

Arnold, B. C., Castillo, E. and Sarabia, J. M. (1993). A variation on the conditional specification theme. *Bulletin of the International Statistical Institute*, **49**, 51–52.

Arnold, B. C., Castillo, E. and Sarabia, J. M. (1994). A conditional characterization of the multivariate normal distribution. *Statistics and Probability Letters*, **19**, 313–315.

Arnold, B. C., Castillo, E. and Sarabia, J. M. (1999). *Conditional Specification of Statistical Models*. Springer-Verlag, New York.

Arnold, B. C., Castillo, E. and Sarabia, J. M. (2002). Conditionally specified multivariate skewed distributions. *Sankhyā, Series A*, **64**, 206–226.

Arnold, B. C. and Wesolowski, J. (1996). Multivariate distributions with Gaussian conditional structure. *Stochastic Processes and Functional Analysis*, 45–49.

Azzalini, A. (1985). A class of distributions which includes the normal ones. *Scandinavian Journal of Statistics*, **12**, 171–178.

Azzalini, A. and Capitanio, A. (1999). Statistical applications of the multivariate skew normal distribution. *Journal of the Royal Statistical Society, Series B*, **61**, 579–602.

Bhattacharyya, A. (1943). On some sets of sufficient conditions leading to the normal bivariate distribution. *Sankhyā*, **6**, 399–406.

Bryc, W. and Plucinska, A. (1985). A characterization of infinite Gaussian sequences by conditional moments. *Sankhyā, Series A*, **47**, 166–173.

Mardia, K. V. (1962). Multivariate Pareto distributions. *Annals of Mathematical Statistics*, **33**, 1008–1015.

Nguyen, T. T., Rempala, G. and Wesolowski, J. (1996). Non-Gaussian measures with Gaussian structure. *Probability and Mathematical Statistics*, **16**, 287–298.

Wesolowski, J. (1991). Gaussian conditional structure of the second order and the Kagan classification of multivariate distributions. *Journal of Multivariate Analysis*, **39**, 79–86.

Chapter 11

Bivariate Matsumoto–Yor Property and Related Characterizations

KONSTANCJA BOBECKA AND JACEK WESOŁOWSKI
*Wydział Matematyki i Nauk Informacyjnych,
Politechnika Warszawska, Warszawa, Poland*

CONTENTS

ABSTRACT

In this paper, a bivariate version of the Matsumoto–Yor independence property for the generalized inverse Gaussian (GIG) and gamma distributions is considered. It appears that the property does not characterize the general families of bivariate gamma and GIG distributions but only special cases of random vectors with either independent or linearly dependent components.

KEYWORDS AND PHRASES: (Bivariate) gamma distribution, generalized inverse Gaussian distribution, Matsumoto–Yor property

11.1 INTRODUCTION

Considering functionals of the geometric Brownian motion, Matsumoto and Yor (2001) have recently observed that the map $\psi : (0, \infty)^2 \to (0, \infty)^2$, defined by $\psi(x, y) = ((x+y)^{-1}, x^{-1} - (x+y)^{-1})$, preserves a probability measure which is a product of the generalized inverse Gaussian (GIG) and the gamma distributions. Recall that the GIG distribution $\mu_{-p,a,b}$ is defined by

$$\mu_{-p,a,b}(dx) = K_1 x^{-p-1} \exp\left(-a^{-1}x - (bx)^{-1}\right)I_{(0,\infty)}(x)\, dx,$$

where $p \in \mathbf{R}$, $a, b \in (0, \infty)$, are the parameters. The gamma distribution $\gamma_{q,c}$ is defined by

$$\gamma_{q,c}(dy) = K_2 y^{q-1} \exp\left(-c^{-1}y\right)I_{(0,\infty)}(y)\, dy,$$

where $q, c \in (0, \infty)$ are parameters and K_1 and K_2 are normalizing constants. Matsumoto and Yor (2001) observed that if random variables X and Y are independent, X has the GIG distribution $\mu_{-p,a,a}$ $(p > 0)$, and Y has the gamma distribution $\gamma_{p,a}$, i.e., $(X, Y) \sim \mu_{-p,a,a} \otimes \gamma_{p,a}$, then the random vector

$$(U, V) = \psi(X, Y) = \left(\frac{1}{X + Y}, \frac{1}{X} - \frac{1}{X + Y}\right)$$

has the same distribution as (X, Y); hence, in particular, U and V are independent. As observed in Letac and Wesołowski (2000), the following extension of the Matsumoto–Yor property holds: if (X, Y) has the distribution $\mu_{-p,a,b} \otimes \gamma_{p,a}$, then (U, V) is distributed according to $\mu_{-p,b,a} \otimes \gamma_{p,b}$.

Matsumoto and Yor (2001) asked about a converse of their observation: Assume that X and Y are independent and that the random vector $(U, V) = \psi(X, Y)$ has independent components. Does (X, Y) have the distribution $\mu_{-p,a,b} \otimes \gamma_{p,a}$ (and consequently (U, V) is distributed according to $\mu_{-p,b,a} \otimes \gamma_{p,b}$)? This question has been answered in the affirmative by Letac and Wesołowski (2000) [a related problem involving constancy of regression of V or V^{-1} on U has been considered also in Seshadri and Wesołowski (2001) and solved finally in Wesołowski (2002)]. Also in that paper, the authors considered the Matsumoto–Yor property for distributions on the cone of positive definite symmetric matrices. The characterization given there was restricted to distributions having strictly positive, twice continuously differentiable densities. An extension assuming differentiable densities has been given in Wesołowski (2002). The problem for random matrices of different dimensions has been studied recently in Massam and Wesołowski (2004).

The Matsumoto–Yor property has never been treated up to now for random vectors. This paper is intended to partially fill this gap by considering the bivariate situation. It is interesting to note that the development of studies here is parallel to investigations concerning the Lukacs (1955) characterization of the gamma law: if X, Y are independent positive non-degenerate random variables and $X + Y$, $X/(X + Y)$ are also independent, then X, Y have gamma distributions. It was followed by the solution of the problem in the matrix variate case first—see Olkin and Rubin (1962), Casalis and Letac (1996), Letac and Massam (1998), and Bobecka and Wesołowski (2002). The case of random vectors was treated only recently, first, bivariate in Bobecka (2002), and then, n-variate in Bobecka and Wesołowski (2004).

11.2   CHARACTERIZATION

Below we present the characterization related to the Matsumoto–Yor property for bivariate random vectors. It appears that in this case the independence property (similarly as in the Lukacs characterization) imposes special structures of the bivariate gamma and GIG distributions. This is the main result of the paper presented in the theorem below.

THEOREM 11.2.1 *Let $\bar{X} = (X_1, X_2)$ and $\bar{Y} = (Y_1, Y_2)$ be independent random vectors with positive components. Assume that $\bar{X}$ or $\bar{Y}$ is not degenerate to the point. Let*

$$\bar{U} = (U_1, U_2) = \left(\frac{1}{X_1 + Y_1}, \frac{1}{X_2 + Y_2} \right)$$

and

$$\bar{V} = (V_1, V_2) = \left(\frac{1}{X_1} - \frac{1}{X_1 + Y_1}, \frac{1}{X_2} - \frac{1}{X_2 + Y_2} \right).$$

The random vectors $\bar{U}$ and $\bar{V}$ are independent if and only if there exist positive constants p_j, λ_j, κ_j, such that X_j has a GIG distribution: $\mu_{-p_j,\lambda_j,\kappa_j}$ and Y_j has a gamma distribution: γ_{p_j,λ_j}, $j = 1, 2$, and either

1. *the components of $\bar{X}$ and $\bar{Y}$ are independent*

or

2. *the components of $\bar{X}$ and $\bar{Y}$ are linearly dependent: $X_1 = aX_2$, $Y_1 = aY_2$ with $a = \lambda_1/\lambda_2$, and then $p_1 = p_2$.*

PROOF. Observe that if any one of $\bar{X}$ and $\bar{Y}$ is not degenerate to a point, then all four random vectors $\bar{X}, \bar{Y}, \bar{U}$, and $\bar{V}$ are not degenerate.

Necessity. The independence property and the identity

$$\frac{Y_j}{X_j} = \frac{V_j}{U_j}, \quad j = 1, 2,$$

imply

$$E\left(Y_1^\alpha Y_2^\beta e^{\sigma_1 Y_1 + \sigma_2 Y_2}\right) E\left(X_1^{-\alpha} X_2^{-\beta} A(\sigma_1, \sigma_2, \theta_1, \theta_2)\right)$$
$$= E\left(V_1^\alpha V_2^\beta e^{\theta_1 V_1 + \theta_2 V_2}\right) E\left(U_1^{-\alpha} U_2^{-\beta} B(\sigma_1, \sigma_2, \theta_1, \theta_2)\right),$$
$$(11.2.1)$$

where

$$A(\sigma_1, \sigma_2, \theta_1, \theta_2) = e^{\sigma_1 X_1 + \sigma_2 X_2 + \theta_1 X_1^{-1} + \theta_2 X_2^{-1}}$$

and

$$B(\sigma_1, \sigma_2, \theta_1, \theta_2) = e^{\sigma_1 U_1^{-1} + \sigma_2 U_2^{-1} + \theta_1 U_1 + \theta_2 U_2}$$

for any negative $\sigma_1, \sigma_2, \theta_1, \theta_2$ and fixed non-negative α and β.

Taking the logarithm of both sides of (11.2.1) and applying $\partial^2/\partial\sigma_1\partial\theta_1$, we obtain

$$\frac{E\left(X_1^{-\alpha+1} X_2^{-\beta} A(\sigma_1, \sigma_2, \theta_1, \theta_2)\right) E\left(X_1^{-\alpha-1} X_2^{-\beta} A(\sigma_1, \sigma_2, \theta_1, \theta_2)\right)}{\left[E\left(X_1^{-\alpha} X_2^{-\beta} A(\sigma_1, \sigma_2, \theta_1, \theta_2)\right)\right]^2}$$

$$= \frac{E\left(U_1^{-\alpha+1} U_2^{-\beta} B(\sigma_1, \sigma_2, \theta_1, \theta_2)\right) E\left(U_1^{-\alpha-1} U_2^{-\beta} B(\sigma_1, \sigma_2, \theta_1, \theta_2)\right)}{\left[E\left(U_1^{-\alpha} U_2^{-\beta} B(\sigma_1, \sigma_2, \theta_1, \theta_2)\right)\right]^2}.$$

$$(11.2.2)$$

Now applying (11.2.1) for α, $\alpha - 1$, and $\alpha + 1$ to (11.2.2), we arrive at

$$\frac{E\left(Y_1^{\alpha-1}Y_2^{\beta}e^{\sigma_1 Y_1 + \sigma_2 Y_2}\right) E\left(Y_1^{\alpha+1}Y_2^{\beta}e^{\sigma_1 Y_1 + \sigma_2 Y_2}\right)}{\left[E\left(Y_1^{\alpha}Y_2^{\beta}e^{\sigma_1 Y_1 + \sigma_2 Y_2}\right)\right]^2}$$

$$= \frac{E\left(V_1^{\alpha-1}V_2^{\beta}e^{\theta_1 V_1 + \theta_2 V_2}\right) E\left(V_1^{\alpha+1}V_2^{\beta}e^{\theta_1 V_1 + \theta_2 V_2}\right)}{\left[E\left(V_1^{\alpha}V_2^{\beta}e^{\theta_1 V_1 + \theta_2 V_2}\right)\right]^2}.$$

$$(11.2.3)$$

Similarly, we obtain a dual relation

$$\frac{E\left(Y_1^{\alpha}Y_2^{\beta-1}e^{\sigma_1 Y_1 + \sigma_2 Y_2}\right) E\left(Y_1^{\alpha}Y_2^{\beta+1}e^{\sigma_1 Y_1 + \sigma_2 Y_2}\right)}{\left[E\left(Y_1^{\alpha}Y_2^{\beta}e^{\sigma_1 Y_1 + \sigma_2 Y_2}\right)\right]^2}$$

$$= \frac{E\left(V_1^{\alpha}V_2^{\beta-1}e^{\theta_1 V_1 + \theta_2 V_2}\right) E\left(V_1^{\alpha}V_2^{\beta+1}e^{\theta_1 V_1 + \theta_2 V_2}\right)}{\left[E\left(V_1^{\alpha}V_2^{\beta}e^{\theta_1 V_1 + \theta_2 V_2}\right)\right]^2}.$$

$$(11.2.4)$$

Writing (11.2.3) for $\alpha = 1$, $\beta = 0$ and writing (11.2.4) for $\alpha = 0$, $\beta = 1$, we have

$$\frac{E\left(Y_j^2 e^{\sigma_1 Y_1 + \sigma_2 Y_2}\right) E\left(e^{\sigma_1 Y_1 + \sigma_2 Y_2}\right)}{\left[E\left(Y_j e^{\sigma_1 Y_1 + \sigma_2 Y_2}\right)\right]^2}$$

$$= \frac{E\left(V_j^2 e^{\theta_1 V_1 + \theta_2 V_2}\right) E\left(e^{\theta_1 V_1 + \theta_2 V_2}\right)}{\left[E\left(V_j e^{\theta_1 V_1 + \theta_2 V_2}\right)\right]^2}$$

$$(11.2.5)$$

for $j = 1, 2$. Then by the principle of separation of variables, (11.2.5) implies

$$\frac{\frac{\partial^2 f}{\partial \sigma_j^2} f}{\left(\frac{\partial f}{\partial \sigma_j}\right)^2} = c_j, \qquad \frac{\frac{\partial^2 g}{\partial \theta_j^2} g}{\left(\frac{\partial g}{\partial \theta_j}\right)^2} = c_j, \quad j = 1, 2, \qquad (11.2.6)$$

where f and g are the Laplace transforms of $\bar{Y}$ and $\bar{V}$, respectively, and c_1, c_2 are some constants greater than one. Then as in Bobecka (2003), we conclude that only the following two cases are possible: either

1. $c_1 \neq c_2$ and then

$$f(\sigma_1, \sigma_2) = (1 - \lambda_1 \sigma_1)^{-p_1} (1 - \lambda_2 \sigma_2)^{-p_2},$$

$$(\sigma_1, \sigma_2) \in (-\infty, \lambda_1^{-1}) \times (-\infty, \lambda_2^{-1})$$

and

$$g(\theta_1, \theta_2) = (1 - \kappa_1 \theta_1)^{-p_1} (1 - \kappa_2 \sigma_2)^{-p_2},$$

$$(\theta_1, \theta_2) \in (-\infty, \kappa_1^{-1}) \times (-\infty, \kappa_2^{-1}),$$

where $p_j = 1/(c_j - 1) > 0$, and $\lambda_j > 0$, $\kappa_j > 0$, $j = 1, 2$, i.e., the random vectors $\bar{Y} = (Y_1, Y_2)$ and $\bar{V} = (V_1, V_2)$ have independent gamma components: $Y_j \sim \gamma_{p_j, \lambda_j^{-1}}$, $V_j \sim \gamma_{p_j, \kappa_j^{1}}$, $j = 1, 2$;

or

2. $c_1 = c_2 = c$ and then

$$f(\sigma_1, \sigma_2) = (1 - \lambda_1 \sigma_1 - \lambda_2 \sigma_2 + \lambda_3 \sigma_1 \sigma_2)^{-p},$$

$$\lambda_1 \sigma_1 + \lambda_2 \sigma_2 - \lambda_3 \sigma_1 \sigma_2 < 1, \qquad (11.2.7)$$

and

$$g(\theta_1, \theta_2) = (1 - \kappa_1 \theta_1 - \kappa_2 \theta_2 + \kappa_3 \theta_1 \theta_2)^{-p},$$

$$\kappa_1 \theta_1 + \kappa_2 \theta_2 - \kappa_3 \theta_1 \theta_2 < 1, \qquad (11.2.8)$$

where $p = 1/(c - 1) > 0$ and $\lambda_1, \lambda_2 > 0$, $\lambda_1 \lambda_2 \geq \lambda_3 \geq 0$, $\kappa_1, \kappa_2 > 0$, $\kappa_1 \kappa_2 \geq \kappa_3 \geq 0$, i.e., the random vectors $\bar{Y}, \bar{V}$ have bivariate gamma distributions.

In the next step of the proof, it will be shown that in the above case 2 we have either $\lambda_3 = \lambda_1 \lambda_2$ and $\kappa_3 = \kappa_1 \kappa_2$, which implies that the components of $\bar{Y}$ and $\bar{V}$ are independent, or $\lambda_3 = 0$ and $\kappa_3 = 0$, which implies that the components of $\bar{Y}$ and $\bar{V}$ are linearly dependent gamma variables.

Again, we apply the principle of separation of variables to (11.2.3) with $\alpha = \beta = 1$ and $\sigma_2 = 0$, arriving at

$$E\left(Y_2 e^{\sigma_1 Y_1}\right) E\left(Y_1^2 Y_2 e^{\sigma_1 Y_1}\right) = d\left[E\left(Y_1 Y_2 e^{\sigma_1 Y_1}\right)\right]^2, \qquad (11.2.9)$$

where $d > 1$ is a constant. Now introduce a new random variable Z with the distribution defined by

$$P_Z(dy_1) = \frac{\int_0^\infty y_2 F(dy_1, dy_2)}{E(Y_2)},$$

where F is the df of $\bar{Y}$ and the integral in the numerator is with respect to y_2. Then after dividing both sides of (11.2.9) by $[E(Y_2)]^2$, we have

$$E\left(e^{\sigma_1 Z}\right) E\left(Z^2 e^{\sigma_1 Z}\right) = d\left[E\left(Z e^{\sigma_1 Z}\right)\right]^2,$$

which means that Z is a gamma random variable, $\gamma_{q,1/\alpha}$. Then in particular

$$E\left(e^{\sigma_1 Z}\right) = \frac{1}{(1-\alpha\sigma_1)^q}. \qquad (11.2.10)$$

Now observe that

$$E\left(Y_2 e^{\sigma_1 Y_1}\right) = E\left(e^{\sigma_1 Z}\right) E\left(Y_2\right).$$

Using the fact that $\bar{Y}$ has the bivariate gamma distribution (with the Laplace transform (11.2.7)) and (11.2.10), we obtain the equation

$$(\lambda_2 - \lambda_3\sigma_1)(1-\alpha\sigma_1)^q = \lambda_2(1-\lambda_1\sigma_1)^{p+1} \qquad (11.2.11)$$

for any $\sigma_1 < \lambda_1^{-1}$. Letting $\sigma_1 \uparrow \lambda_1^{-1}$, it follows that the right-hand side of (11.2.11) tends to zero. Consequently, either $\lambda_3 = \lambda_1\lambda_2$ or $\alpha = \lambda_1$. Thus, in the first case we have

$$(1-\alpha\sigma_1)^q = (1-\lambda_1\sigma_1)^p,$$

which implies $\alpha = \lambda_1$ and $q = p$. In the second case, it follows that

$$(\lambda_2 - \lambda_3\sigma_1) = \lambda_2(1-\lambda_1\sigma_1)^{p+1-q}$$

and, thus, either $\lambda_3 = 0$ and then $p+1 = q$ or $\lambda_3 \neq 0$ and then $q = p$, $\lambda_1\lambda_2 = \lambda_3$.

Summing up, only the following cases are possible: either $\lambda_3 = \lambda_1\lambda_2$ or $\lambda_3 = 0$. Similarly, we can show that either $\kappa_3 = \kappa_1\kappa_2$ or $\kappa_3 = 0$.

If $\lambda_3 = \lambda_1\lambda_2$ and $\kappa_3 = \kappa_1\kappa_2$, then $\bar{Y}$ and $\bar{V}$ have independent gamma components: $Y_j \sim \gamma_{p,\lambda_j^{-1}}$, $V_j \sim \gamma_{p,\kappa_j^{-1}}$, $j = 1, 2$.

If $\lambda_3 = 0$ and $\upsilon_3 = 0$, then the components of $\bar{Y}$ and $\bar{V}$ are linearly dependent, i.e., $Y_2 = aY_1$, $V_2 = bV_1$, where $Y_1 \sim \gamma_{p,\lambda_1^{-1}}$, $V_1 \sim \gamma_{p,\kappa_1^{-1}}$, $a = \lambda_2/\lambda_1$, $b = \kappa_2/\kappa_1$.

Observe that other cases are impossible. If $\lambda_3 = \lambda_1\lambda_2$ and $\kappa_3 = 0$, then (Y_1, Y_2) has a density and (V_1, V_2) doesn't have a density. However, if (Y_1, Y_2) has a density, then also $(X_1 + Y_1, X_2 + Y_2) = (U_1, U_2)$ has a density. Hence, $(U_1 + V_1, U_2 + V_2) = \left(\frac{1}{X_1}, \frac{1}{X_1}\right)$ has a density. Thus,

$$(V_1, V_2) = \left(\frac{1}{X_1} - \frac{1}{X_1 + Y_1}, \frac{1}{X_2} - \frac{1}{X_2 + Y_2}\right)$$

has also a density since it is a smooth function of the random vector $(\bar{X}, \bar{Y})$ with independent bivariate absolutely continuous components $\bar{X}$ and $\bar{Y}$. Consequently, $\kappa_3 \neq 0$. Similarly, the case $\lambda_3 = 0$ and $\kappa_3 = \kappa_1 \kappa_2$ is impossible.

Summing up, we have the following two cases:

either

    1. $\bar{Y}$ and $\bar{V}$ have independent gamma components: $Y_j \sim \gamma_{p_j, \lambda_j^{-1}}, V_j \sim \gamma_{p_j, \kappa_j^{-1}}, j = 1, 2,$

or

    2. $\bar{Y}$ and $\bar{V}$ have linearly dependent gamma components: $Y_2 = aY_1, V_2 = bV_1$, where $Y_1 \sim \gamma_{p, \lambda_1^{-1}}, V_1 \sim \gamma_{p, \kappa_1^{-1}}$, $a = \lambda_2/\lambda_1, b = \kappa_2/\kappa_1.$

Case 1

In this case all the random vectors $\bar{X}, \bar{Y}, \bar{U}, \bar{V}$ have densities. Since $\bar{X}, \bar{Y}$ are independent and $\bar{U}, \bar{V}$ are independent, we have the following identity for the densities:

$$f_{\bar{U}}(u_1, u_2) f_{\bar{V}}(v_1, v_2)$$

$$= \frac{f_{\bar{X}}\left(\frac{1}{u_1 + v_1}, \frac{1}{u_2 + v_2}\right) f_{\bar{Y}}\left(\frac{1}{u_1} - \frac{1}{u_1 + v_1}, \frac{1}{u_2} - \frac{1}{u_2 + v_2}\right)}{(u_1 + v_1)^2 u_1^2 (u_2 + v_2)^2 u_2^2}, \quad (11.2.12)$$

which holds a.e. with respect to the Lebesgue measure L_4 in $\mathbf{R}^4$ for $u_j, v_j \in (0, \infty), j = 1, 2$. Using the fact that $\bar{Y}$ and $\bar{V}$ have independent gamma components, we obtain the following:

$$f_{\bar{U}}(u_1, u_2) u_1^{p_1 + 1} u_2^{p_2 + 1} e^{\kappa_1^{-1} u_1} e^{\kappa_2^{-1} u_2} e^{\lambda_1^{-1} u_1^{-1}} e^{\lambda_2^{-1} u_2^{-1}}$$

$$= c f_{\bar{X}}\left((u_1 + v_1)^{-1}, (u_2 + v_2)^{-1}\right) (u_1 + v_1)^{-(p_1 + 1)} (u_2 + v_2)^{-(p_2 + 1)}$$

$$\times e^{\kappa_1^{-1}(u_1 + v_1)} e^{\kappa_2^{-1}(u_2 + v_2)} e^{\lambda_1^{-1}(u_1 + v_1)^{-1}} e^{\lambda_2^{-1}(u_2 + v_2)^{-1}}, \quad (11.2.13)$$

for $u_j, v_j \in (0, \infty), j = 1, 2, L_4$ a.e., where $c = const.$

Denoting $u_1 + v_1 = m_1$, $u_2 + v_2 = m_2$, the above equation can be written as

$$f_{\bar{U}}(u_1, u_2) = c(m_1, m_2) g_1(u_1) g_2(u_2), \tag{11.2.14}$$

where c is the right-hand side of (11.2.13) and

$$g_j(u_j) = u_j^{-p_j-1} e^{-\kappa_j^{-1} u_j - \lambda_j^{-1} u_j^{-1}},$$

$j = 1, 2$. We can always choose m_1, m_2 such that (11.2.14) holds for $(u_1, u_2) \in (0, m_1) \times (0, m_2)$ L_2 a.e. Moreover, m_1 and m_2 can be chosen arbitrarily large. This implies that $\bar{U}$ has independent GIG components $U_j \sim \mu_{-p_j, \kappa_j, \lambda_j}$, $j = 1, 2$. Dually, by (11.2.13), it follows that $\bar{X}$ has also independent GIG components $X_j \sim \mu_{-p_j, \lambda_j, \kappa_j}$, $j = 1, 2$.

Case 2

Since $Y_2 = aY_1$, $V_2 = bV_1$ P-a.s. and $V_j = \frac{1}{X_j} - \frac{1}{X_j + Y_j}$, $j = 1, 2$, we obtain

$$\frac{bY_1}{X_1(X_1 + Y_1)} = \frac{aY_1}{X_2(X_2 + aY_1)} \quad P - \text{a.s.}$$

Since Y_1 is P-a.s. positive, we obtain

$$Y_1(X_1 - bX_2) = X_1^2 - \frac{b}{a} X_2^2 \quad P - \text{a.s.} \tag{11.2.15}$$

Assume now that $X_1 \neq bX_2$ on a set A of positive probability P. Then on A we have

$$Y_1 = \frac{X_1^2 - \frac{b}{a} X_2^2}{X_1 - bX_2},$$

which contradicts the independence of $\bar{X}$ and $\bar{Y}$. Thus, $X_1 = bX_2$ P-a.s. and by (11.2.15) $b = 1/a$. Thus, the components of $\bar{X}$ are linearly dependent: $X_2 = aX_1$. Since $U_j = \frac{1}{X_j + Y_j}$, $j = 1, 2$, we obtain immediately that the components of $\bar{U}$ are also linearly dependent: $U_2 = bU_1$.

Thus, the problem is reduced to the univariate case. Hence, by the result of Letac and Wesołowski (2000), we get that X_1 and U_1 have GIG distributions: $X_1 \sim \mu_{-p, \lambda_1, \kappa_1}$, $U_1 \sim \mu_{-p, \kappa_1, \lambda_1}$.

Sufficiency. Now we assume that the random vectors $\bar{X}$ and $\bar{Y}$ have the GIG and gamma distributions as given in the statement of the theorem. We will show that the random vectors $\bar{U}$ and $\bar{V}$ are independent.

First consider the case of independent components of $\bar{X}$ and $\bar{Y}$. Since, by the assumption of the theorem, $\bar{X}$ and $\bar{Y}$ are independent, it follows that the random vectors (X_1, Y_1) and (X_2, Y_2) are independent. It implies that the random vectors (U_1, V_1) and (U_2, V_2) are independent. However, by the univariate Matsumoto–Yor property (recall that the components of $\bar{X}$ are GIGs and the components of $\bar{Y}$ are gammas), it follows that U_1, V_1 are independent and U_2, V_2 are independent. Again, using the independence of (U_1, V_1) and (U_2, V_2), we conclude that $\bar{U} = (U_1, U_2)$ and $\bar{V} = (V_1, V_2)$ are independent.

Finally, consider the case of linearly dependent components of $\bar{U}$ and $\bar{V}$, i.e., $\bar{U} = (U_1, U_1/a)$ and $\bar{V} = (V_1, V_1/a)$, with U_1 being a GIG random variable and V_1 being a gamma random variable. Then, by the univariate Matsumoto–Yor property, it follows that U_1 and V_1 are independent. Consequently, $\bar{U}$ and $\bar{V}$ are also independent. ∎

REFERENCES

Bobecka, K. (2002). Regression versions of Lukacs type characterizations for the bivariate gamma distribution. *Journal of Applied Statistical Science*, **11**, 213–233.

Bobecka, K. and Wesołowski, J. (2002). The Lukacs–Olkin–Rubin theorem without invariance of the "quotient." *Studia Math.*, **152**, 147–160.

Bobecka, K. and Wesołowski, J. (2004). Multivariate Lukacs theorem. *Journal of Multivariate Analysis*, **91**, 143–160.

Casalis, M. and Letac, G. (1996). The Lukacs–Olkin–Rubin characterization of Wishart distributions on symmetric cones. *Annals of Statistics*, **24**, 763–786.

Letac, G. and Massam, H. (1998). Quadratic and inverse regressions for Wishart distributions. *Annals of Statistics*, **26**, 573–595.

Letac, G. and Wesołowski, J. (2000). An independence property for the product of GIG and gamma laws. *Annals of Probability*, **28**, 1371–1383.

Lukacs, E. (1955). A characterization of the gamma distribution. *Annals of Mathematical Statistics*, **26**, 319–324.

Massam, H. and Wesołowski, J. (2003). The Matsumoto–Yor property and the structure of the Wishart distribution. *Journal of Multivariate Analysis*, to appear.

Matsumoto, H. and Yor, M. (2001). An analogue of Pitman's $2M - X$ theorem for exponential Wiener functionals. Part II: The role of the generalized inverse Gaussian laws. *Nagoya Mathematical Journal*, **162**, 65–86.

Matsumoto, H. and Yor, M. (2003). Interpretation via Brownian motion of some independence properties between GIG and gamma variables. *Statistics & Probability Letters*, **61**, 253–259.

Olkin, I. and Rubin, H. (1962). A characterization of the Wishart distribution. *Annals of Mathematical Statistics*, **33**, 1272–1280.

Seshadri, V. and Wesołowski, J. (2001). Mutual characterizations of the gamma and the generalized inverse Gaussian laws by constancy of regression. *Sankhyā, Series A*, **63**, 107–112.

Wesołowski, J. (2002). The Matsumoto–Yor independence property for GIG and Gamma laws, revisited. *Mathematical Proceedings of the Cambridge Philosophical Society*, **133**, 153–161.

Chapter 12

First Principal Component Characterization of a Continuous Random Variable

CARLES M. CUADRAS
Department of Statistics, University of Barcelona,
Barcelona, Spain

CONTENTS

ABSTRACT

A random variable with continuous distribution is expanded as a series of principal components. Some properties of the first principal component, which may characterize the variable, and an inequality concerning a function and its derivative are obtained. The logistic distributions have special interest, as the first principal component is the cumulative distribution function. Dependence between variables is also studied.

KEYWORDS AND PHRASES: Orthogonal expansions, principal components, Karhunen–Loève expansion, logistic distribution, stochastic dependence

12.1 INTRODUCTION

Let X be a continuous random variable with range $I = [a, b]$, cdf F, and density f with respect to the Lebesgue measure. We can relate X to the stochastic process $\mathbf{X} = \{X_t, t \in I\}$, where X_t is the indicator of $[X > t]$. Then $X_t^2 = X_t$ and it can be proved that the expansion is

$$|X - X'| = \int_I (X_t - X_t')^2 \, dt = \sum_{n \geq 1} \{f_n(X) - f_n(X')\}^2,$$

where X' is distributed as X, and if a is finite

$$X = a + \int_I X_t \, dt = a + \sum_{n \geq 1} f_n(b) f_n(X),$$

$$X = a + \int_I X_t^2 \, dt = a + \sum_{n \geq 1} f_n(X)^2.$$

The sequence $\{f_n(X)\}$ is a countable set of uncorrelated random variables, principal components of $\mathbf{X}$, with variances $\mathrm{Var}(f_n(X)) = \lambda_n$, such that

$$\mathrm{tr}(K) = \int_I F(x)\{1 - F(x)\} \, dx = \sum_{n \geq 1} \lambda_n,$$

λ_n being an eigenvalue with eigenfunction ψ_n of the integral operator defined by the symmetric kernel $K(s, t) = \min\{F(s), F(t)\} - F(s)F(t)$.

The set $\{\psi_n\}$ of eigenfunctions constitutes a basis of $L^2([a, b])$ and each function f_n is obtained by

$$f_n(x) = \int_I X_t \psi_n(t) \, dt = \int_a^x \psi_n(s) \, ds.$$

As a consequence of Mercer's theorem, the above expansions exist provided that tr(K) is finite. These expansions can be obtained from the Karhunen–Loève expansion of **X**

$$X_t = \sum_{n \geq 1} \psi_n(t) X_n. \tag{12.1.1}$$

The convergence is in the mean square sense. See Cuadras and Fortiana (1995, 2000) and Cuadras and Lahlou (2000).

It is worth noting that zero correlation between principal components can result from the generalized Hoeffding's formula

Cov $(\alpha(X), \beta(Y))$

$$= \int_a^b \int_c^d \{H(x, y) - F(x)G(y)\} \, d\beta(y) \, d\alpha(x), \tag{12.1.2}$$

where H is a bivariate cdf with univariate marginals F, G, and ranges $[a, b]$, $[c, d]$, by taking $H - FG = K$, $\alpha = f_m$, $\beta = f_n$ (Cuadras, 2002a).

The orthogonal expansion of a random variable in principal components is of interest in formulating a continuous extension of multidimensional scaling (Cuadras and Fortiana, 1995), in obtaining a graphical test to distinguish between logistic and normal distributions (Cuadras and Lahlou, 2000 and Cuadras and Cuadras, 2002), and in improving some tests of independence by relating principal components (Cuadras, 2002b), and the eigenvalues of K contribute to the study of the asymptotic distribution of some statistics related to Rao's quadratic entropy (Liu and Rao, 1995). For tests of fit, Durbin and Knott (1972) used a similar principal components expansion for $\sqrt{n}(F_n - F)$, where F_n is the empirical cdf based on a sample of size n obtained from X.

This paper aims to characterize the distribution of X via the first principal component of **X**.

12.2   THE DIFFERENTIAL EQUATION

Let $y_n = f_n$ and $\mu_n = E(f_n(X))$. It can be proved that the means μ_n, variances λ_n, and functions y_n satisfy the second-order differential equation

$$\lambda y'' + (y - \mu)f = 0, \qquad y(a) = y'(a) = 0. \qquad (12.2.1)$$

The solution of this equation is well known when X is $[0, 1]$ uniform. The solutions for X exponential, logistic and Pareto were obtained by Cuadras and Fortiana (1995) and Cuadras and Lahlou (2000, 2002), respectively.

Examples of principal components $f_n(X)$ and the corresponding variances λ_n are:

1. $(\sqrt{2}/(n\pi))(1 - \cos n\pi X)$, $\lambda_n = 1/(n\pi)^2$, if X is $[0, 1]$ uniform.
2. $\left[2J_0(\xi_n \exp(-X/2)) - 2J_0(\xi_n)\right]/\xi_n J_0(\xi_n)$, $\lambda_n = 4/\xi_n^2$, if X is exponential with unit mean, where ξ_n is the n-th positive root of J_1 and J_0, J_1 are the Bessel functions of the first order.
3. $(n(n+1))^{-1/2}[L_n(F(X)) + (-1)^{n+1}\sqrt{2n+1}]$, $\lambda_n = 1/\{n(n+1)\}$, if X is standard logistic, where (L_n) are the Legendre polynomials on $[0, 1]$.
4. $c_n[X\sin(\xi_n/X) - \sin(\xi_n)]$, $\lambda_n = 3/\xi_n^2$, if X is Pareto with $F(x) = 1 - x^{-3}$, $x > 1$, where $c_n = 2\xi_n^{-1/2}(2\xi_n - \sin\{2\xi_n\})^{-1/2}$ and $\xi_n = \tan(\xi_n)$.

12.3 SOME PROPERTIES OF THE EIGENFUNCTIONS

In this section we study some properties of the eigenfunctions ψ_n and their integrals f_n.

PROPOSITION 12.3.1 *The first eigenfunction ψ_1 is strictly positive and satisfies*

$$\psi_1(x) > \psi_n(a) = \psi_n(b) = 0, \quad x \in (a, b), \quad n \geq 1.$$

PROOF. K is positive, so ψ_1 is also positive (Perron-Frobenius theorem). On the other hand $K(t, t) = F(t)(1 - F(t)) = \sum_{n \geq 1} \lambda_n \psi_n(t)^2$, which satisfies $K(a, a) = K(b, b) = 0$. ∎

Clearly f_1 is increasing and positive. Moreover, $\text{tr}(K)$ is finite if Var (X) exists and any $f_n(b)$ is bounded even for $b = \infty$.

PROPOSITION 12.3.2 *Let* $\sigma^2 = Var(X)$. *Then* $f_n(b)$ *satisfies*

$$|f_n(b)| < \frac{\sigma}{\sqrt{\lambda_n}}, \quad n \geq 1. \qquad (12.3.1)$$

PROOF. $\psi_n = f_n'$ is an eigenfunction and from (12.1.2)

$$\int_I \left(\int_I K(x,y)\psi_n(x)\,dx \right) dy = \lambda_n \int_I \psi_n(y)\,dy$$
$$= \lambda_n f_n(b)$$
$$= \text{Cov}\,(f_n(X), X).$$

Hence $\lambda_n^2 f_n(b)^2 < \lambda_n \sigma^2$. ∎

PROPOSITION 12.3.3 *The principal components* $\{f_n(X)\}$ *of* **X** *constitutes a complete orthogonal system of* $L^2(F)$.

PROOF. The orthogonality can be proved as a consequence of (12.1.2). Let $\phi \in L^2(F)$. The proof that $\{f_n(X)\}$ is complete is easy if we assume that ϕ' exists. Suppose that $\text{Cov}\,(\phi(X), f_n(X)) = 0$, $n \geq 1$. As $\{\psi_n\}$ is a complete system $\phi' = \sum_{n \geq 1} c_n \psi_n$ and integrating, we have $\phi = c_0 + \sum_{n \geq 1} c_n f_n$. But $\text{Cov}\,(\phi(X), f_n(X)) = c_n \lambda_n = 0$, $n \geq 1$, which shows that ϕ must be constant. ∎

12.4 THE FIRST PRINCIPAL COMPONENT

In this section we prove two interesting properties of the first principal component $f_1(X)$.

PROPOSITION 12.4.1 *The increasing function* f_1 *characterizes the distribution of* X.

PROOF. Write $y = f_1$. Then y satisfies the differential equation (12.2.1), where $\mu = E(y(X))$, $\lambda = \text{Var}\,(y(X))$. When the function y is given, μ and λ may be obtained by solving the equations

$$\int_I \frac{-\lambda y''}{(y - \mu)}\,dx = 1, \qquad \int_I \frac{-\lambda y''}{(y - \mu)} y\,dx = \mu.$$

Then the density of X is given by $f = -\lambda y''/(y - \mu)$. ∎

PROPOSITION 12.4.2 *Let $\rho^2(X_t, \phi(X))$ denote the squared correlation between X_t and a function $\phi(X)$. The average of $\rho^2(X_t, \phi(X))$ weighted by $K(t,t) = F(t)(1 - F(t))$ is maximum for $\phi \equiv f_1$, i.e.,*

$$\sup_{\phi} \int_I \rho^2(X_t, \phi(X)) K(t,t)\, dt = \int_I \rho^2(X_t, f_1(X)) K(t,t)\, dt.$$

PROOF. Let us write (see Proposition 12.3.3) $\phi = \sum_{n \geq 1} a_n f_n$. Then $\text{Var}\,(\phi(X)) = \sum_{n \geq 1} a_n^2 \lambda_n$ and we can suppose $\sum_{n \geq 1} a_n^2 = 1$. From (12.1.1)

$$\int_I \text{Cov}\,(X_t, \phi(X))^2 dt = \sum_{n \geq 1} a_n^2 \lambda_n^2.$$

As $\text{Var}\,(X_t) = K(t,t)$, we have

$$\sup_{\phi} \int_I \rho^2(X_t, \phi(X)) K(t,t)\, dt = \left(\sum_{n \geq 1} a_n^2 \lambda_n^2 \right) \Big/ \left(\sum_{n \geq 1} a_n^2 \lambda_n \right)$$

$$\leq \lambda_1 \left(\sum_{n \geq 1} a_n^2 \lambda_n \right) \Big/ \left(\sum_{n \geq 1} a_n^2 \lambda_n \right)$$

$$= \lambda_1. \quad \blacksquare$$

12.5 AN INEQUALITY

The following inequality holds for X with the normal $N(0,1)$ distribution [Chernoff (1981) and Cacoullos (1982)]

$$[E(\phi'(X))]^2 \leq \text{Var}\,[\phi(X)] \leq E([\phi'(X)]^2),$$

where ϕ is an absolutely continuous function and $\phi(X)$ has finite variance. This inequality was extended to any distribution by Klaassen (1985). Let us prove a related inequality concerning the function of a random variable and its derivative.

If $f_1(X)$ is the first principal dimension, then $\psi_1 = f_1'$ and $f_1(b) = \int_I \psi_1(x)\, dx$. Noting that $f_1(b)$ is bounded, let us define the probability density with support $I = [a, b]$

$$\varphi(y) = \frac{\psi_1(y)}{f_1(b)}.$$

THEOREM 12.5.1 *Let Y be a r.v. with pdf φ. If ϕ is an absolutely continuous function and $\phi(X)$ has finite variance then the following inequality holds*

$$\text{Var}\,[\phi(X)] \geq f_1(b)^2\,\text{Var}\,[f_1(X)][E(\phi'(Y))]^2, \qquad (12.5.1)$$

with equality if ϕ is f_1.

PROOF. From Proposition 12.3.3, we can write $\phi' = \sum_{n\geq 1} a_n \psi_n$, where $a_1 = \int_I \psi_1(x)\phi'(x)dx = f_1(b)E(\phi'(Y))$. Then $\phi = \sum_{n\geq 1} a_n f_n$ and

$$\text{Var}\,[\phi(X)] = \sum_{n\geq 1} a_n^2\,\text{Var}\,(f_n(X)) \geq a_1^2\,\text{Var}\,(f_1(X)).$$

If $\phi \equiv f_1$ we have $f_1(b)^2[E(\phi'(Y))]^2 = 1$. ∎

12.6 THE LOGISTIC DISTRIBUTION

Suppose that X follows the standard logistic distribution. The cdf is

$$F(x) = (1 + \exp(-x))^{-1}, \quad -\infty < x < +\infty,$$

and the density is $f = F(1 - F)$. This distribution has special interest, as the first and the second principal components are directly related to F and f.

The two first principal components are $f_1 = \sqrt{6}F$, $f_2 = \sqrt{30}f$, i.e., proportional to the cdf and the density, respectively [Cuadras and Lahlou (2000)]. Note that f_1 can be obtained directly, as if we write $f_1 = cF$, then $\mu = c/2$, $\lambda = c^2/12$ and (12.2.1) reduces to

$$\frac{c^2}{12}(1 - 2F) + \left(F - \frac{1}{2}\right) = 0,$$

so $c = \sqrt{6}$. Besides

$$\sup_{\phi} \int_I \rho^2(X_t, \phi(X))f(t)\,dt = \int_I \rho^2(X_t, F(X))f(t)\,dt,$$

i.e., the expectation of $\rho^2(X_t, \phi(X))$ with respect to $f(t)$ is maximum for $F(X)$.

As now the above density $\varphi = \psi_1/h_1(b)$ is f and $f_1(b) = \sqrt{6}$, inequality (12.5.1) for the logistic distribution reduces to

$$\text{Var}\,[\phi(X)] \geq 3[E(\phi'(X))]^2.$$

In general, if Z is logistic with variance σ^2 then $Z = \alpha X$ with $\alpha = (\sqrt{3}/\pi)\sigma$. Noting that the functions f_n', $n \geq 2$, are orthogonal to $f_1' = \sqrt{6}f$ and using (12.1.2), we obtain

$$\text{Cov}\,(F(X), \phi(\alpha X)) = \int_I (\sqrt{6}f(x))^2\, dx \int_I \frac{\alpha}{2} f(y)\phi'(\alpha y)\, dy$$

$$= \frac{\alpha}{2} E(\phi'(\alpha X)).$$

As $\text{Var}\,(F(X)) = 1/12$, the Cauchy-Schwarz inequality proves that

$$\text{Var}\,[\phi(Z)] \geq \left(\frac{3^2}{\pi^2}\right) \text{Var}\,(Z)[E(\phi'(Z))]^2.$$

12.7   DEGREE OF INDEPENDENCE

The principal components can be used to measure the degree of independence between two random variables X, Y, with ranges $[a, b]$, $[c, d]$ joint density h and marginal densities f, g, respectively.

In an early paper, Cuadras (1972) defined independence of degree k between X, Y as

$$E(X^i Y^j) = E(X^i)E(Y^j) \quad \text{for} \quad i + j \leq k + 1,$$

provided that the moments exist, where i, j, k are positive integers. Clearly X and Y are uncorrelated if $k = 1$ and stochastically independent if $k = \infty$.

An extension is as follows. Let $\{f_m(X)\}$ and $\{g_n(Y)\}$ be the principal components for X and Y, respectively, ordered according to the corresponding eigenvalues. Independence of degree k is now defined as

$$E\{f_i(X)g_j(Y)\} = E\{f_i(X))E(g_j(Y)\} \quad \text{for} \quad i + j \leq k + 1.$$

If $k = 1$ then X, Y are uncorrelated in an extended sense. For example, if the marginal distributions are logistic, the first principal components f_1, g_1 are the cumulative distribution functions F, G and Spearman's rho is

$$\rho_S = \text{Corr } (F(X), G(Y)) = 0.$$

On the other hand, as a consequence of the following theorem, X, Y are stochastically independent if $k = \infty$.

THEOREM 12.7.1 *Let X, Y be r.v.'s with supports $[a, b], [c, d]$, joint cdf H, and marginals F, G, respectively. If $\{f_m(X)\}, \{g_n(Y)\}$ are the principal components, then X, Y are stochastically independent if and only if all the correlations between components are zero:*

$$\text{Corr } (f_n(X), g_m(Y)) = 0 \quad \text{for} \quad m, n \geq 1.$$

PROOF. We have $f_m(x) = \int_a^x \psi_m(t) \, dt, g_n(y) = \int_c^y \phi_n(t) \, dt$, where $\{\psi_m\}, \{\phi_n\}$ are complete orthonormal sets. Then $\{\psi_m \times \phi_n\}$ is a complete orthonormal set on $L^2([a, b] \times [c, d])$. As $H - F G \in L^2([a, b] \times [c, d])$ we may expand

$$H(x, y) - F(x)G(y) = \sum_{m,n \geq 1} a_{mn} \psi_m(x) \phi_n(y).$$

The Fourier coefficients are

$$\begin{aligned} a_{mn} &= \int_a^b \int_c^d \{H(x, y) - F(x)G(y)\} \psi_m(x) \phi_n(y) \, dxdy \\ &= \int_a^b \int_c^d \{H(x, y) - F(x)G(y)\} \, dh_m(x) \, dg_n(y) \\ &= \text{Cov } \{h_n(X), g_m(Y)\} \quad \text{(from (12.1.2)).} \end{aligned}$$

Hence, $H(x, y) = F(x)G(y)$ if and only if $a_{mn} = 0$ for $m, n \geq 1$.

An example of independence of order $1 < k < \infty$ is as follows. Let $F_i = (f_i - \mu_i)/\sqrt{\lambda_i}$, i.e., $F_i(X)$ is the standardized principal component and, likewise, consider G_i. Let us define the bivariate probability density h with marginals f, g:

$$h(x, y) = f(x)g(y) \left[1 + \sum_{i=k}^m \rho_i F_i(x) G_i(y) \right], \quad 1 \leq k \leq m,$$

where $|\rho_i| < 1, i = k, \ldots, m$, play the role of canonical correlations (Cuadras, 2002b). It is readily proved that independence of order k exists for this distribution. ∎

REFERENCES

Cacoullos, T. (1982). On upper and lower bounds for the variance of a function of a random variable. *Annals of Probability*, **10**, 799–809.

Chernoff, H. (1981). A note on an inequality involving the normal distribution. *Annals of Probability*, **9**, 533–535.

Cuadras, C. M. (1972). Theoretical, experimental foundations and new models of factor analysis. *Investigación Pesquera,* **36**, 163–169 (in Spanish).

Cuadras, C. M. (2002a). On the covariance between functions. *Journal of Multivariate Analysis*, **81**, 19–27.

Cuadras, C. M. (2002b). Diagonal distributions via orthogonal expansions and tests of independence. In *Distributions with Given Marginals and Statistical Modelling* (Eds., C. M. Cuadras, J. Fortiana, and J. A. Rodriguez-Lallena), pp. 35–42, Kluwer Academic Press, Dordrecht, The Netherlands.

Cuadras, C. M. and Cuadras, D. (2002). Orthogonal expansions and distinction between logistic and normal. In *Goodness-of-fit Tests and Model Validity* (Eds., C. Huber-Carol, N. Balakrishnan, M. S. Nikulin, and M. Mesbah), pp. 327–339, Birkhäuser, Boston, Massachusetts.

Cuadras, C. M. and Fortiana, J. (1995). A continuous metric scaling solution for a random variable. *Journal of Multivariate Analysis*, **52**, 1–14.

Cuadras, C. M. and Fortiana, J. (2000). The importance of geometry in multivariate analysis and some applications. In *Statistics for the 21st Century* (Eds., C. R. Rao and G. Szekely), pp. 93–108, Marcel Dekker, New York.

Cuadras, C. M. and Lahlou, Y. (2000). Some orthogonal expansions for the logistic distribution. *Communications in Statistics— Theory and Methods,* **29**, 2643–2663.

Cuadras, C. M. and Lahlou, Y. (2002). Principal components of the Pareto distribution. In *Distributions with Given Marginals and Statistical Modelling* (Eds., C. M. Cuadras, J. Fortiana, and J. A. Rodriguez-Lallena), pp. 43–50, Kluwer Academic Press, Dordrecht, The Netherlands.

Durbin, J. and Knott, M. (1972). Components of Cramér-von Mises statistics. I. *Journal of the Royal Statistical Society, Series B*, **34**, 290–307.

Klaasen, C. A. J. (1985). On an inequality of Chernoff. *Annals of Probability,* **13**, 966–974.

Liu, Z. and Rao, C. R. (1995). Asymptotic distribution of statistics based on quadratic entropy and bootstrapping. *Journal of Statistical Planning and Inference,* **43**, 1–18.

Chapter 13

The Lawless–Wang's Operational Ridge Regression Estimator under the LINEX Loss Function

ESRA AKDENIZ AND FIKRI AKDENIZ
Department of Statistics, University of Cukurova,
Adana, Turkey

CONTENTS

ABSTRACT

In this paper, using the asymmetric LINEX loss function we derive the risk function of the operational ridge regression estimator (RRE) for individual regression coefficients. We also examine the risk performance of this estimator when the LINEX loss function is used.

KEYWORDS AND PHRASES: LINEX loss function, multicollinearity, operational estimator, ridge regression, risk function

13.1 INTRODUCTION

In regression analysis, researchers often encounter the problem of multicollinearity. The ridge regression estimators proposed by Hoerl and Kennard (1970) are one of the solutions to solve the problem of multicollinearity. This estimator, however, is non-operational since it depends upon a biasing parameter, k, which is unknown. A particular operational version of the estimator, where k is estimated from the data and is, therefore, stochastic, has been proposed by Lawless and Wang (1976). However, as the expressions for the first two moments of the ridge regression estimator are complicated, it is not easy to carry out numerical evaluations. Since the work of Hoerl and Kennard (1970), many studies on small sample properties of the ridge regression estimators have been made. Firinguetti (1987) derived the exact bias and matrix of second-order moments of the Lawless and Wang operational ridge regression estimator. Firinguetti (1991) derived the exact properties of the Lawless and Wang operational ridge regression estimator. Kozumi and Ohtani (1994) derived the general expressions for the moments of the ordinary ridge regression coefficients in a different way from Firinguetti (1987). It is interesting to note that almost all studies on the ridge regression and biased estimators use the mean square error or, equivalently, the (symmetric) quadratic loss as the basis of measuring estimators' performance. Being symmetric, the quadratic loss imposes equal penalty on over- and underestimation of the same magnitude. Symmetric losses, such as squared error loss, are widely employed in decision theory, but their application is often justified by their nice mathematical properties, not their appropriateness in representing a true loss structure. It is well known that the use of symmetric loss functions may be inappropriate in many circumstances, particularly when positive and negative errors have different consequences. Why should the ordinary least squares estimator be used to measure the performance of another estimator when the loss function is not the squared error loss function? The nature of many decision problems, such as reliability analysis, requires the use of asymmetric losses. For example, when estimating the average life of the components of a spaceship, overestimation is usually more serious than underestimation. In fact, the space shuttle disaster of 1986 was partly the result of the management's overestimation of the mean life or reliability of the solid-fuel rocket booster [see Feynman (1987)]. In some estimation problems it may be appropriate to use asymmetric loss function. Varian (1975) introduced a very useful asymmetric loss function called a LINEX loss function. Zellner (1986) extensively discussed the properties of the LINEX loss function. Zellner (1986) also suggests that in dam construction, underestimation of the peak water level is often more serious than overestimation.

When estimating the parameter θ, by $\hat{\theta}$, the loss function is given by:

$$L(\hat{\theta}) = b \left[\exp(a\Delta) - a\Delta - 1 \right] \tag{13.1.1}$$

where $a \neq 0$ is a shape parameter, $b > 0$ is a factor of proportionality, and $\Delta = (\hat{\theta} - \theta)/\theta$ is the relative estimation error in using $\hat{\theta}$ to θ estimate.

Since the relative estimation error does not depend on a unit, it is often used. In our investigation we assume (without loss of generality) that $b = 1$. The sign of shape parameter a reflects the direction of asymmetry and the magnitude of a reflects the degree of asymmetry. There are numerous practical applications where we use a loss function such as (13.1.1). When $a > 0$, the convex loss (13.1.1) increases almost linearly for negative error Δ and almost exponentially for positive error. Therefore, overestimation is a more serious mistake than underestimation. When $a < 0$, the linear–exponential increases are interchanged, where underestimation is more serious than overestimation. For small values of $|a|$, $L(\hat{\theta}) \doteq ba^2(\hat{\theta} - \theta)^2/2\theta^2$, which is proportional to a squared loss. Thus, the LINEX loss function can be regarded as a generalization of the squared error loss function allowing for asymmetry and we employ an asymmetric LINEX loss function as an error criterion.

LINEX has been extensively explored in the literature and found to be quite useful. Numerous authors have considered the LINEX loss function in various problems of interest. Examples are Rojo (1987), Giles and Giles (1996), Zou (1997), Pandey (1997), Ohtani (1999), Wan and Kurumai (1999), Takada (2000), Xiao (2000), Wen and Levy (2001), and Sarabia and Pascual (2002). In particular, Ohtani (1995) considered the risk of the feasible generalized ridge regression (FGRR) estimator under the LINEX loss function. Wan (1999) examined the properties of the feasible, almost unbiased generalized ridge regression estimator under the asymmetric LINEX loss function.

This article examines the properties of Lawless–Wang's operational ridge regression estimator under the asymmetric LINEX loss function. In Section 13.2, the model and estimators are presented. In Section 13.3, the risk function and risk performance of Lawless–Wang's operational ridge regression estimator have been given.

13.2   THE MODEL AND ESTIMATORS

Consider the classical linear regression model (CLRM)

$$y = Z\gamma + u \tag{13.2.1}$$

where y is a vector of observations in the dependent variable, Z is an $n \times l$ full rank matrix of non-stochastic observations in the explanatory variables, γ is an $l \times 1$ vector of unknown coefficients, and u is an $n \times 1$ vector of unobserved random disturbances such that $u \sim N(0, \sigma^2 I)$. Let us define Λ as the diagonal matrix of eigenvalues and Q as the corresponding matrix of orthonormal eigenvectors of $Z'Z$. Then we can rewrite (13.2.1) as follows:

$$y = ZQQ'\gamma + u = X\beta + u, \tag{13.2.2}$$

with

$$X = ZQ, \qquad X'X = \Lambda, \quad \text{and} \quad \beta = Q'\gamma. \tag{13.2.3}$$

Equation (13.2.2) is the canonical version of the CLRM. Thus, the least squares estimator of β is

$$b = \Lambda^{-1}X'y, \tag{13.2.4}$$

whose ith element is simply

$$b_i = \frac{x_i'y}{\lambda_i}, \quad i = 1, 2, \dots, l \tag{13.2.5}$$

where x_i is the ith column of X and λ_i is the ith diagonal element of Λ. It is well known that for this estimator, we have

$$E(b) = \beta, \quad \text{Var}(b) = \sigma^2\Lambda^{-1}, \quad \text{Var}(b_i) = \sigma^2/\lambda_i. \tag{13.2.6}$$

It is obvious that b_i becomes inefficient as the value of λ_i gets small. However, for ill-conditioned Z matrices, Hoerl and Kennard (1970) proposed the ordinary ridge regression (ORR) estimator defined in the orthogonal model (13.2.2) to be

$$\hat{\beta} = (\Lambda + kI)^{-1}\Lambda b, \quad k > 0. \tag{13.2.7}$$

Denoting the ith element of $\hat{\beta}$ as $\hat{\beta}_i$, $\hat{\beta}_i$ is written as

$$\hat{\beta}_i = \frac{\lambda_i}{\lambda_i + k}b_i.$$

Generally, k is a function of the data and the properties of the estimator will depend on the choice of k. As for the ORR estimator, Lawless and Wang (1976) proposed, as the biasing parameter, to use

$$\hat{k} = \frac{l\hat{\sigma}^2}{b'\Lambda b} \tag{13.2.8}$$

where $\hat{\sigma}^2 = (y - Xb)'(y - Xb)/(n-l)$. Thus, the ORR estimator proposed by Lawless and Wang is written as

$$\tilde{\beta} = \left(\Lambda + \frac{l\hat{\sigma}^2}{b'\Lambda b}I\right)^{-1}\Lambda b \tag{13.2.9}$$

$$= \left(I + \frac{l\hat{\sigma}^2}{b'\Lambda b}\Lambda^{-1}\right)^{-1}b,$$

and the ith element of $\tilde{\beta}$ is given by

$$\tilde{\beta}_i = \left(1 - \frac{l\hat{\sigma}^2}{\lambda_i b'\Lambda b + l\hat{\sigma}^2}\right)b_i. \tag{13.2.10}$$

From the simulation results of Lawless and Wang (1976), it can be found that the ORR estimator (13.2.9) has the best mean square error (MSE) properties among other ridge regression estimators. Therefore, we confine ourselves to the ORR estimator proposed by Lawless and Wang (1976) hereafter.

13.3   RISK PERFORMANCE OF $\tilde{\beta}_i$ UNDER THE LINEX LOSS FUNCTION

The asymmetric LINEX loss function of b_i is defined as

$$
L(b_i) = \exp(a\Delta) - a\Delta - 1
$$
$$
= \sum_{r=2}^{\infty} \frac{a^r \Delta_i^r}{r!} \tag{13.3.1}
$$

where $\Delta_i = (b_i - \beta_i)/\beta_i$. The risk function of b_i is

$$
R(b_i) = E(L(b_i)) = \exp\left(\frac{a^2}{2\theta_i^2}\right) - 1 \tag{13.3.2}
$$

where $\theta_i = \lambda_i^{1/2}\beta_i/\sigma$. The risk of Lawless–Wang's estimator can be written as

$$
R(\tilde{\beta}) = E(L(\tilde{\beta}_i)) = E(\exp(a\tilde{\Delta}_i) - a\tilde{\Delta}_i - 1)
$$

where $\tilde{\Delta}_i = (\tilde{\beta}_i - \beta_i)/\beta_i$. Using (13.2.10), we have

$$
[(\tilde{\beta}_i - \beta_i)/\beta_i]^r = \left[\left(1 - \frac{l\hat{\sigma}^2}{\lambda_i b'\Lambda b + l\hat{\sigma}^2}\right)\frac{z_i}{\theta_i} - 1\right]^r \tag{13.3.3}
$$
$$
= \sum_{j=0}^{r} {}_rC_j \left[\left(1 - \frac{l\hat{\sigma}^2}{\lambda_i b'\Lambda b + l\hat{\sigma}^2}\right)^j\right]\left(\frac{z_i}{\theta_i}\right)^j (-1)^{r-j}
$$
$$
= \sum_{j=0}^{r}\sum_{p=0}^{j} {}_rC_j\, {}_jC_p (-1)^{r+j-p}
$$
$$
\times \left[\left(\frac{l\hat{\sigma}^2}{\lambda_i b'\Lambda b + l\hat{\sigma}^2}\right)^p\right]\left(\frac{z_i}{\theta_i}\right)^j
$$

where $z_i = \lambda_i^{1/2} b_i / \sigma$, $_mC_r = \frac{m!}{r!(m-r)!}$, and $\frac{b_i}{\beta_i} = \frac{z_i}{\theta_i}$. The risk function of $\tilde{\beta}_i$ is

$$R(\tilde{\beta}_i) = E(L(\tilde{\beta}_i)) = \sum_{r=2}^{\infty} \frac{a^r}{r!} E \left(\frac{\tilde{\beta}_i - \beta_i}{\beta_i} \right)^r \qquad (13.3.4)$$

$$= \sum_{r=2}^{\infty} a^r \sum_{j=0}^{r} \sum_{p=0}^{j} \frac{(-1)^{r+j-p}}{(r-j)! p! (j-p)!}$$

$$\times E \left[\left(\frac{l\hat{\sigma}^2}{\lambda_i b' \Lambda b + l\hat{\sigma}^2} \right)^p \right] \left(\frac{z_i}{\theta_i} \right)^j .$$

Define the functions $H(p,q)$ and $J(p,q)$ as

$$H(p,q) = E \left[\left(\frac{l\hat{\sigma}^2}{\lambda_i b' \Lambda b + l\hat{\sigma}^2} \right)^p \left(\frac{z_i}{\theta_i} \right)^{2q} \right] \qquad (13.3.5)$$

$$J(p,q) = E \left[\left(\frac{l\hat{\sigma}^2}{\lambda_i b' \Lambda b + l\hat{\sigma}^2} \right)^p \left(\frac{z_i}{\theta_i} \right)^{2q+1} \right] \qquad (13.3.6)$$

where p and q are nonnegative integers. The explicit formulas for $H(p,q)$ and $J(p,q)$ are

$$H(p,q) = \sum_{s=0}^{\infty} \sum_{t=0}^{\infty} K'_{st} \int_0^1 \frac{z^{q+s+t+\frac{l^*-1}{2}} (1-z)^{p+\frac{\nu}{2}-1}}{\left[1 + \left(\frac{\lambda_i}{c} - 1 \right) z \right]^p} \, dz \quad (13.3.7)$$

if $j = 2q$ where

$$K'_{st} = \frac{1}{\left(\frac{\theta_i^2}{2} \right)^q} \eta_s(\omega_1) \eta_t(\omega_2)$$

$$\times \frac{\Gamma(q+s+t+n/2)\Gamma(q+s+1/2)}{\Gamma(\nu/2)\Gamma(s+1/2)\Gamma(q+s+t+(l^*+1)/2)}$$

$$\eta_s(\omega_1) = \frac{e^{-\omega_1/2}(\omega_1/2)^s}{s!}, \qquad \eta_t(\omega_2) = \frac{e^{-\omega_2/2}(\omega_2/2)^t}{t!}.$$

or

$$J(p,q) = \sum_{s=0}^{\infty} \sum_{t=0}^{\infty} \tilde{K}_{st} \int_0^1 \frac{z^{q+s+t+\frac{l^*}{2}} (1-z)^{p+\frac{\nu}{2}-1}}{\left[1 + \left(\frac{\lambda_i}{c} - 1 \right) z \right]^p} \, dz \quad (13.3.8)$$

if $j = 2q + 1$ where

$$\tilde{K}_{st} = \frac{1}{\left(\frac{\theta_i^2}{2}\right)^{q+1/2}} \eta_s(\omega_1)\eta_t(\omega_2)$$

$$\times \frac{\Gamma(q + s + t + (n+1)/2)\Gamma(q + s + 1)}{\Gamma(\nu/2)\Gamma(s + 1/2)\Gamma(q + s + t + l^*/2)}.$$

In comparing the risks, we consider relative efficiency of Lawless–Wang's operational ridge regression estimator to the OLS estimator, defined by

$$\frac{R(b_i)}{R(\tilde{\beta}_i)}.$$

Note that $\tilde{\beta}_i$ has smaller risk than b_i if the relative risk $\frac{R(b_i)}{R(\tilde{\beta}_i)} > 1$ and vice versa. See the Appendix for the derivations of $H(p, q)$ and $J(p, q)$.

APPENDIX

Since

$$\left(\frac{l\hat{\sigma}^2}{\lambda_i b' \Lambda b + l\hat{\sigma}^2}\right) = \frac{l\hat{\sigma}^2/\sigma^2}{\lambda_i(\lambda_i b_i^2/\sigma^2) + \sum_{u=1, u \neq i}^l (\lambda_u b_u^2/\sigma^2) + l\hat{\sigma}^2/\sigma^2}$$

$$= \frac{cV_3}{\lambda_i V_1 + \lambda_i V_2 + cV_3}$$

we have

$$H(p, q) = E\left[\left(\frac{cV_3}{\lambda_i V_1 + \lambda_i V_2 + cV_3}\right)^p \left(\frac{z_i}{\theta_i}\right)^{2q}\right]$$

$$= \frac{1}{\theta_i^{2q}} E\left[\left(\frac{cV_3}{\lambda_i V_1 + \lambda_i V_2 + cV_3}\right)^p V_1^p\right]$$

where $c = l/\nu$, $b \sim N(\beta_i, \sigma^2/\lambda_i)$, $z_i^2 = V_1$, $V_1 = \lambda_i b_i^2/\sigma^2 \sim \chi_1^2(\omega_1)$, and $\chi_1^2(\omega_1)$ is the non-central chi-square distribution with 1 degree of freedom and non-centrality parameter $\omega_1 = \lambda_i \beta_i^2/\sigma^2$. Also, we see that $V_2 = \sum_{u=1, u \neq i}^l (\lambda_u b_u^2/\sigma^2) \sim \chi_{l^*}^2(\omega_2)$, $l^* = l - 1$ and $\omega_2 = \sum_{u=1, u \neq i}^l (\lambda_u \beta_u^2/\sigma^2)$, $V_3 = \nu\hat{\sigma}^2/\sigma^2 \sim \chi_\nu^2$, where χ_ν^2 is the central chi-square distribution with $\nu = n - l$ degrees of freedom, and V_1, V_2, V_3 are mutually independent. Using V_1, V_2, V_3 and $c = l/v$, $H(p,q)$ is

$$\frac{1}{\theta_i^{2q}} E\left[\left(\frac{cV_3}{\lambda_i V_1 + \lambda_i V_2 + cV_3}\right)^p V_1^p\right]$$

$$= \sum_{s=0}^{\infty} \sum_{t=0}^{\infty} K_{st} \int_0^{\infty} \int_0^{\infty} \int_0^{\infty} V_1^{q+s-(1/2)} V_2^{t+(l^*/2)-1} V_3^{(v/2)-1}$$

$$\times \left(\frac{cV_3}{\lambda_i V_1 + \lambda_i V_2 + cV_3}\right)^p \exp\left(-\frac{V_1 + V_2 + V_3}{2}\right)$$

$$\times dV_3 dV_2 dV_1, \tag{A.1}$$

where

$$K_{st} = \frac{1}{\theta_i^{2q}} \eta_s(\omega_1)\eta_t(\omega_2)$$

$$\times \frac{1}{2^{v/2}\Gamma(v/2)2^{s+1/2}\Gamma(s+1/2)2^{t+l^*/2}\Gamma(t+l^*/2)},$$

$$\eta_s(\omega_1) = \frac{e^{-\omega_1/2}(\omega_1/2)^s}{s!},$$

$$\eta_t(\omega_2) = \frac{e^{-\omega_2/2}(\omega_2/2)^t}{t!}. \tag{A.2}$$

Making use of the change of variables $t_1 = V_1/V_3$, $t_2 = V_2/V_3$, and $t_3 = V_3$, (A.1) reduces to

$$= \sum_{s=0}^{\infty} \sum_{t=0}^{\infty} K_{st} \int_0^{\infty} \int_0^{\infty} \int_0^{\infty} t_1^{q+s-(1/2)} t_2^{t+(l^*/2)-1} t_3^{q+s+t+(l^*+v-1)/2}$$

$$\times \left(\frac{c}{\lambda_i t_1 + \lambda_i t_2 + c}\right)^p \exp\left[-\frac{(t_1 + t_2 + 1)t_3}{2}\right] dt_3 dt_2 dt_1 \tag{A.3}$$

where the Jacobian is t_3^2. Making use of the change of variable, $u_1 = (t_1 + t_2 + 1)t_3/2$, and recognizing that

$$\int_0^\infty t_3^{q+s+t+(l^*+v-1)/2} \exp\left[-\frac{(t_1+t_2+1)t_3}{2}\right] dt_3$$

$$= \int_0^\infty \left(\frac{2u_1}{t_1+t_2+1}\right)^{q+s+t+(l^*+v-1)/2}$$

$$\times \exp(-u_1)\left(\frac{2}{t_1+t_2+1}\right) du_1$$

$$= \frac{2^x}{(t_1+t_2+1)^x}\int_0^\infty u_1^{x-1}$$

$$\times \exp(-u_1)\, du_1 = \frac{2^x}{(t_1+t_2+1)^x}\Gamma(x)$$

where $x = q + s + t + (l^* + v + 1)/2$. Hence, (A.3) reduces to

$$= \sum_{s=0}^\infty \sum_{t=0}^\infty K'_{st} \int_0^\infty \int_0^\infty t_1^{q+s-(1/2)} t_2^{t+(l^*/2)-1}$$

$$\times \left(\frac{c}{\lambda_i t_1 + \lambda_i t_2 + c}\right)^p (t_1+t_2+1)^{-x} dt_2 dt_1 \qquad (A.4)$$

where $K'_{st} = K_{st} 2^x \Gamma(x)$. Again, making use of the change of variables $u_2 = t_1 + t_2$ and $u_3 = t_2/(t_1+t_2)$, (A.4) reduces to

$$\sum_{s=0}^\infty \sum_{t=0}^\infty K'_{st} \int_0^\infty \int_0^1 (1-u_3)^{q+s-1/2} u_2^{q+s+t+(l^*-1)/2}$$

$$\times (1+u_2)^{-x} u_3^{t+l^*/2-1}\left(\frac{c}{\lambda_i u_2 + c}\right)^p du_3 du_2.$$

Recognizing that

$$\int_0^1 (1-u_3)^{q+s+1/2-1} u_3^{t+l^*/2-1} du_3 = \frac{\Gamma(t+l^*/2)\Gamma(q+s+1/2)}{\Gamma(q+s+t+(l^*+1)/2)},$$

we have

$$= \sum_{s=0}^\infty \sum_{t=0}^\infty K''_{st} \int_0^\infty u_2^y (1+u_2)^{-x}\left(\frac{c}{\lambda_i u_2 + c}\right)^p du_2, \qquad (A.5)$$

where

$$y = q + s + t + (l^* - 1)/2$$

and

$$K''_{st} = K'_{st} \frac{\Gamma(t + l^*/2)\Gamma(q + s + 1/2)}{\Gamma(q + s + t + (l^* + 1)/2)}.$$

Finally, making use of the change of variable $z = u_2/(1 + u_2)$, (A.5) reduces to

$$= \sum_{s=0}^{\infty} \sum_{t=0}^{\infty} K''_{st} \int_0^1 \left(1 + \frac{z}{1-z}\right)^{-x} \left(\frac{z}{1-z}\right)^y \qquad (A.6)$$

$$\times \left(\frac{c}{\lambda_i \frac{z}{1-z} + c}\right)^p \frac{1}{(1-z)^2} dz$$

$$= \sum_{s=0}^{\infty} \sum_{t=0}^{\infty} K''_{st} \int_0^1 \frac{z^{q+s+t+(l^*-1)/2}(1-z)^{p+v/2-1}}{\left[1 + \left(\frac{\lambda_i}{c} - 1\right)z\right]^p} dz.$$

Using (A.7), we obtain $H(p, q)$ given in (13.3.7). Then we obtain the risk function of $\tilde{\beta}_i$ given in (13.3.4). The corresponding expression for $J(p, q)$ given in (13.3.8) can be obtained in a parallel way and is left to the readers to verify.

REFERENCES

Feynman, R. P. (1987). "Mr Feynman goes to Washington". *Engineering and Science*. Fall 1987, California Institute of Technology, 6–62.

Firinguetti, L. (1987). Exact moments of Lawless and Wang's operational ridge regression estimator. *Communications in Statistics—Theory and Methods*, **16**, 731–745.

Giles, J. A. and Giles, D. E. A. (1996). Estimation of the regression scale after a pre-test for homoscedasticity under the LINEX loss. *Journal of Statistical Planning Inference*, **50**, 21–35.

Hoerl, A. E. and Kennard, R. W. (1970). Ridge regression: Biased estimation for non-orthogonal problems. *Technometrics*, **12**, 55–67.

Kozumi, H. and Ohtani, K. (1994). The general expressions for the moments of Lawless and Wang's ordinary ridge regression estimator. *Communications in Statistics—Theory and Methods*, **23**, 2755–2774.

Lawless, J. F. and Wang, P. (1976). A simulation study of ridge and other regression estimators. *Communications in Statistics—Theory and Methods*, **5**, 307–323.

Ohtani, K. (1995). Generalized ridge regression estimator under the LINEX loss function. *Statistical Papers*, **36**, 99–110.

Ohtani, K. (1999). Risk performance of a pre-test estimator for normal variance with the Stein-variance estimator under the LINEX loss function. *Statistical Papers*, **40**, 75–87.

Pandey, B. N. (1997). Testimator of the scale parameter of the exponential distribution using LINEX loss function. *Communications in Statistics—Theory and Methods*, **26**, 2191–2202.

Rojo, J. (1987). On the admissibility of with respect to the LINEX loss function. *Communications in Statistics—Theory and Methods*, **16**, 3745–3748.

Sarabia, J. M. and Pascual, M. (2002). A class of Lorenz curves based on linear exponential loss functions. *Communications in Statistics—Theory and Methods*, **31**, 925–942.

Takada, Y. (2000). Sequential point estimation of normal mean under LINEX loss function. *Metrika*, **52**, 163–171.

Varian, H. R. (1975). A Bayesian approach to real estate assessment. In *Studies of Bayesian Econometrics and Statistics in Honor of Leonard J. Savage* (Eds., S. E. Fienberg and A. Zellner), pp. 195–208, North-Holland, Amsterdam, The Netherlands.

Wan, A. T. K. (1999). A note on almost unbiased generalized ridge estimator under asymmetric loss. *Journal of Statistical Computation and Simulation*, **62**, 411–421.

Wan, A. T. K. and Kurumai, H. (1999). An alternative feasible minimum mean squared error estimator of the disturbance variance in linear regression under asymmetric loss. *Statistics & Probability Letters*, **45**, 253–259.

Wen, D. and Levy, M. S. (2001). BLINEX: A bounded asymmetric loss function with application to Bayesian estimation. *Communications in Statistics—Theory and Methods*, **30**, 147–153.

Xiao, Y. (2000). LINEX unbiasedness in a prediction problem. *Annals of the Institute of Statistical Mathematics*, **52**, 712–721.

Zellner, A. (1986). Bayesian estimation and prediction using asymmetric loss functions. *Journal of the American Statistical Association*, **81**, 446–451.

Zou, G. (1997). Admissible estimation for finite population under the LINEX loss function. *Journal of Statistical Planning and Inference*, **62**, 373–384.

Chapter 14

On the Distribution of the Reference Dose and Its Application in Health Risk Assessment

MEHDI RAZZAGHI

Department of Mathematics, Computer Science and Statistics, Bloomsburg University, Bloomsburg, PA, U.S.A.

CONTENTS

ABSTRACT

Humans are constantly exposed to a multitude of chemicals through inhalation, food, drugs, and other routes. One of the challenges of the regulatory agencies is to determine a safe exposure level for various chemical substances. Accurate human data are seldom available, and to assess toxicity of chemicals, animal bioassay experiments are frequently conducted in controlled environments. One approach in determining safe exposure levels is to first establish a no-observed-adverse-effect-level (NOAEL) and divide it by a series of uncertainty factors to obtain a Reference Dose (RfD) for humans. The NOAEL, therefore, can be viewed as a discrete random variable with a finite support, which is the set of all experimental doses. Although a default value of 10 has traditionally been used for each uncertainty factor, recently it has been shown that uncertainty factors behave as random variables and that their distributions may be approximated by the lognormal distribution.

Using the distributional properties of the NOAEL and also the product of uncertainty factors, here we consider the distribution of the RfD. It is shown that this distribution can be expressed as a mixture of lognormal distributions. The properties of the distribution are discussed and it is demonstrated that the approach can depend on the experimental design and also the number of animals per dose level. These instabilities stem from problems associated with the NOAEL determination. A model approach, based on Benchmark Doses, is recommended.

KEYWORDS AND PHRASES: Distribution, lognormal, risk assessment, reference dose, uncertainty factors

14.1   INTRODUCTION

During the last few decades, recent advances in technological development have resulted in a high rate of production of chemicals. Because of the potential health hazards and possible adverse environmental impacts of these chemicals, there has been a growing interest in the scientific evaluation for the potential risk of these agents. Risk is generally defined as the probability of an adverse effect as the result of exposure to a chemical agent, and risk assessment is a systematic process for describing and quantifying the risk associated with hazardous substances. The process of risk assessment was formally defined in a publication by the National Research Council (NRC, 1983) commonly referred to as the "Red Book." Accordingly, the process of risk assessment is divided into four stages: hazard identification, exposure assessment, dose-response assessment, and quantitative risk characterization. A critical step in the final stage of this process is to determine safe exposure levels for humans. The most conservative approach would be to ban any agent that produces an adverse effect at a certain dose. This approach, however, is unacceptable since any chemical agent could produce toxic effects at high levels. Even many of the essential minerals and metals known to have several curative and preventive effects become toxic at high doses. As pointed out by Flaten (1997), if the dose is high enough, all essential elements are toxic.

Because of unavailability and shortcomings of human data, animal bioassays are frequently conducted to establish the relationship between the incidence of disease and exposure (dose) to a toxicant. Typically, a few dose levels, including unexposed controls, are selected for study in an animal bioassay. To assess the toxicity of a carcinogenic compound, regulatory agencies, such as the Environmental Protection Agency (EPA), have traditionally fit a dose-response model to the bioassay data. The model is then used to obtain an estimate of a Virtually Safe Dose (VSD) which is defined as the dose corresponding to a small negligible increase in risk over that of unexposed animals. For noncancer endpoints, on the other hand, the current approach relies on the determination of a No-Observed-Adverse-Effect-Level (NOAEL), defined as the highest experimental dose level which produces no significant statistical or biological evidence of health effects over the background in a population. In order to obtain a Reference Dose (RfD) for humans, the NOAEL is generally divided by a series of uncertainty factors to account for uncertainties due to using animal data for human effects, sensitivities in a human population, and uncertainty of predicting chronic effects from subchronic studies and others. The RfD for noncancer chemicals can, therefore, be defined as

$$\text{RfD} = \frac{\text{NOAEL}}{U_1 \times U_2 \times \cdots \times U_k}, \qquad (14.1.1)$$

where $U_1, \ldots, U_k$ represent the uncertainty factors. Recently, however, there has been much criticism for using NOAEL for risk assessment. As discussed by Gaylor (1994), the NOAEL-based method is ill defined, is somewhat subjective, is highly dependent on the dosel spacing, and does not reward better experimentation. Leisenring and Ryan (1992) considered the statistical properties of the NOAEL. Using a Weibull dose-response model, they show that the NOAEL will often occur at dose levels associated with substantially increased risks over the controls. Since in practice it is always the RfD that is used for risk assessment, it may be argued that division of NOAEL by the uncertainty factors should remove concerns about instabilities associated with the NOAEL. Here, we consider the distribution of RfD and investigate its statistical properties. In the next section we derive the exact distribution of RfD and discuss its properties in Section 14.3. In Section 14.4, we consider some special cases of dosing regimen, and in Section 14.5, statistical properties of the RfD distribution are investigated through a numerical example.

14.2 DISTRIBUTION OF RfD

In the absence of information to select a specific value for an uncertainty factor, default values of 10 are traditionally used. Dourson, Felter, and Robinson (1996) and Hattis (1998) show, by examining several databases, that uncertainty factors can be considered as random variables and that their distribution can be approximated by the lognormal distribution. Utilizing this result, Kodell and Gaylor (1999) and Gaylor and Kodell (2000) consider the joint distribution of the uncertainty factors and derive upper confidence limits for their combined range.

Suppose that an experiment consists of a control and $g - 1$ non-zero dose levels, $0 = d_1 < d_2 < \cdots < d_g$. Let D denote the experimental dose level at which the NOAEL occurs and suppose that

$$P(D = d_i) = p_i, \qquad i = 1, \ldots, g,$$

$$\sum_{i=1}^{g} p_i = 1 \qquad\qquad (14.2.1)$$

defines the distribution of D. Then, if R is the random variable designating the value of RfD, we have $R = D/U$, where $U = U_1 \times U_2 \times \cdots \times U_k$ is the product of the uncertainty factors. Now, suppose that each U_i $(i = 1, 2, \ldots, g)$ is a lognormally distributed random variable with

$$\mu_i = E(\log U_i),$$
$$\sigma_i^2 = \mathrm{Var}\,(\log U_i), \qquad\qquad (14.2.2)$$

Then, assuming the independence of $U_1, U_2, \ldots, U_k$, clearly U has a lognormal distribution with

$$\mu = E(U) = \sum_{j=1}^{k} \mu_j$$

and

$$\sigma^2 = \mathrm{Var}\,(U) = \sum_{j=1}^{k} \sigma_j^2.$$

Furthermore, we have [see Johnson, Kotz, and Balakrishnan (1994)]

$$\eta = E(U) = \exp\left(\frac{1}{2\sigma^2} - \frac{\mu}{\sigma}\right) \qquad\qquad (14.2.3)$$

and

$$\tau^2 = \text{Var}(U) = \exp\left(\frac{1}{2\sigma^2} - \frac{2\mu}{\sigma^2}\right)\left\{\exp\left(\frac{2}{\sigma^2}\right) - 1\right\}$$

$$(14.2.4)$$

as the mean and variance of U. Moreover, the distribution of R can be derived as

$$G_R(r) = P(R \leq r) = P\left(\frac{D}{U} \leq r\right)$$

$$= \sum_{i=1}^{g} P(D = d_i)P\left(\frac{D}{U} \leq r \mid D = d_i\right)$$

$$= \sum_{i=1}^{g} p_i P\left(\frac{d_i}{U} \leq r\right) = \sum_{i=1}^{g} p_i \left[1 - P\left(U \leq \frac{d_i}{r}\right)\right]$$

$$= \sum_{i=1}^{g} p_i \left[1 - F\left(\frac{d_i}{r}\right)\right], \qquad (14.2.5)$$

where F denotes the cumulative distribution function of a lognormal distribution with mean and variance given by (14.2.3) and (14.2.4), respectively. Also, from (14.2.5), the density of R can be obtained as

$$g_R(r) = \frac{1}{r^2}\sum_{i=1}^{g} p_i d_i \, f\left(\frac{d_i}{r}\right)$$

$$= \sum_{i=1}^{g} p_i \left(\frac{1}{r\sigma\sqrt{2\pi}}\right) \exp\left\{-(\log r - \log d_i + \mu)^2/2\sigma^2\right\}.$$

$$(14.2.6)$$

Equation (14.2.6) shows that the distribution of RfD can be expressed as a finite mixture of g lognormal distributions with the mean and variance of the ith component being $\log d_i - \mu$ and σ^2, respectively, and the mixing proportions being $p_1, p_2, \ldots, p_g$.

14.3   PROPERTIES OF THE RfD DISTRIBUTION

From (14.2.6), it is readily seen that the ℓth moment of R is given by

$$
\begin{aligned}
m_\ell = E(R^\ell) &= \sum_{i=1}^{g} p_i \int_0^\infty \frac{r^{\ell-1}}{\sigma\sqrt{2\pi}} \\
&\quad \times \exp\{-(\log r - \log d_i + \mu)^2/2\sigma^2\}\,dr \\
&= \sum_{i=1}^{g} p_i \exp\{(\log d_i - \mu)\ell + \sigma^2\ell^2/2\} \\
&= \exp(-\mu\ell + \sigma^2\ell^2/2)\sum_{i=1}^{g} p_i d_i^\ell, \quad \ell = 0, 1, 2, \ldots
\end{aligned}
$$

$$(14.3.1)$$

In particular, the mean and variance of R are, respectively, given by

$$
E(R) = e^{-\mu+\sigma^2/2} \sum_{i=1}^{g} p_i d_i \tag{14.3.2}
$$

and

$$
\text{Var}(R) = e^{-2\mu+\sigma^2} \left\{ e^{\sigma^2}\sum_{i=1}^{g} p_i d_i^2 - \left(\sum_{i=1}^{g} p_i d_i\right)^2 \right\}. \tag{14.3.3}
$$

Also, closed-form expressions can be derived for the coefficients of skewness, kurtosis, and variation as [see Johnson, Kotz, and Balakrishnan (1994)]

$$
\gamma_1 = \frac{e^{3\sigma^2} v_3 - 3e^{\sigma^2} v_1 v_2 + 2v_1^3}{[e^{\sigma^2} v_2 - v_1^2]^{3/2}},
$$

$$
\gamma_2 = \frac{e^{6\sigma^2} v_4 - 4e^{3\sigma^3} v_1 v_3 + 6e^{\sigma^2} v_1^2 v_2 - 3v_1^4}{[e^{\sigma^2} v_2 - v_1^2]^2} - 3,
$$

$$
\text{c.v.} = \frac{[e^{\sigma^2} v_2 - v_1^2]^{1/2}}{v_1}
$$

where ν_ℓ, $\ell = 1, 2, 3, \ldots$, is the ℓth moment of D given by

$$\nu_\ell = \sum_{i=1}^{g} p_i d_i^\ell.$$

Note that although (14.3.1) provides all moments of R, simpler approximate expressions can directly be obtained [see Rohatgi (1984, p. 268)] in terms of moments of D and U as

$$m_1 = E(R) \approx \frac{\nu}{\eta} + \frac{\nu \tau^2}{\eta^3}, \tag{14.3.4}$$

$$m_\ell = E(R^\ell) \approx \frac{\nu^\ell}{\eta^\ell} + \frac{\ell}{2} \left\{ (\ell - 1) \frac{\nu^{\ell-1}}{\eta^\ell} \delta^2 + (\ell + 1) \frac{\nu^\ell}{\eta^{\ell+2}} \tau^2 \right\},$$

$$\ell = 2, 3, \ldots, \tag{14.3.5}$$

where

$$\nu = \nu_1 = E(D), \qquad \delta^2 = \nu_2 - \nu_1^2 = \mathrm{Var}\,(D). \tag{14.3.6}$$

Using (14.3.4) and (14.3.5), an approximate expression for variance of R is obtained as

$$\mathrm{Var}\,(R) \approx \frac{\nu^2}{\eta^2} \left[\frac{\delta^2}{\delta^2 + \nu^2} + \frac{\tau^2}{\tau^2 + \eta^2} \right].$$

14.4   SOME SPECIAL CASES

In practice, often the doses for the study are selected according to some criteria. In this section, we consider two such special cases.

14.4.1   Equal spacing

Many bioassay experiments are designed to have equi-spaced doses over the desired range. For example, in toxicological studies, often a maximum tolerated dose (MTD), defined by the National Cancer Institute [see, e.g., Sontag, Page, and Safiotti (1976)] as the "highest dose of a test agent during the chronic study that can be predicted not to alter the animals' normal longevity from effects other than carcinogenicity," is identified and three or four equi-distant dose levels are selected between controls and the MTD.

Suppose, therefore, $d_i = (i - 1)d$ for $i = 1, \ldots, g$, where d is an appropriately chosen constant. Then, from (14.3.1), we have

$$m_\ell = d^\ell e^{-\mu\ell+\sigma^2\ell^2/2} \sum_{i=1}^{g} p_i(i - 1)^\ell, \quad \ell = 0, 1, 2, \ldots.$$

Specifically, the mean and variance of R are, in this case, given by

$$E(R) = d(v - 1)\, e^{\mu+\sigma^2/2}$$

and

$$\mathrm{Var}\,(R) = d^2 e^{-2\mu+\sigma^2}\left\{e^{\sigma^2}(\delta^2 + v^2 - 2v + 1) - (v - 1)^2\right\},$$

respectively, where v and δ are as given in (14.3.6).

14.4.2 Geometric dosing

Another common method of dose selection is to first choose the lowest non-zero dose level, which is generally selected at a value associated with no or little toxicity, such as $LD_{.01}$, defined as a dose level expected to cause toxicity in 1% of animals. This value is then multiplied by a constant $k > 1$ to obtain the next and the subsequent dose levels. Suppose, therefore, $d_i = k^{i-1}d$ for $i = 1, 2, \ldots, g$. Then, from (14.3.1), we have

$$m_\ell = \left(\frac{d}{k}\right)^\ell e^{-\mu\ell+\sigma^2\ell^2/2} \sum_{i=1}^{g} p_i k^{\ell i}, \quad \ell = 0, 1, 2, \ldots \quad (14.4.1)$$

which shows that in this case the moments of R are defined in terms of the probability generating function of D. More specifically, the summation in (14.4.1) is the generating function of the convolution of ℓ independent observations from the NOAEL distribution. If we further assume that the probability distribution of D can be expressed as a truncated geometric distribution,

$$p_i = \frac{1 - p}{1 - p^g}\, p^{i-1}, \quad i = 1, 2, \ldots, g,$$

then (14.4.1) reduces to

$$m_\ell = d^\ell \frac{1 - p}{1 - p^g}\frac{1 - (pk^\ell)^g}{1 - pk^\ell}\, e^{-\mu\ell+\sigma^2 t^2/2}, \quad \ell = 0, 1, 2, \ldots.$$

14.5 NUMERICAL ILLUSTRATION

In order to investigate the properties and behavior of the distribution of RfD, we use the results of a small simulation. In their analysis of the distribution of NOAEL, Leisenring and Ryan (1992) consider a typical safety assessment experiment with a control and three exposure levels scaled to the interval [0, 1], assumed to be 0, 0.01, 0.10, and 1. For any exposure level d, a Weibull model

$$\pi(d) = 1 - \exp\left(-\gamma_0 - \gamma_1 d^\alpha\right), \quad d \geq 0$$
$$= 0, \qquad\qquad\qquad\qquad\text{otherwise} \tag{14.5.1}$$

is used to express the probability of an adverse effect. Because of its flexibility and diversity, the Weibull distribution is very often used as a dose-response model in risk assessment. As α changes, a wide range of models is obtained. Leisenring and Ryan (1992) select the parameter values in (14.5.1) by initially assuming a 0% background response rate and sample sizes of 50. They then find the values that force the probability that the NOAEL takes on the highest experimental dose level to be very small. This leads to $\gamma_1 = 0.35$. Now, by keeping γ_1 fixed, the background dose is changed from 0 to 3 and 10%, which, in turn, results in highest response rates of 32 and 36%, respectively. The probability that NOAEL is equal to each experimental exposure level is then derived for different dose-response shapes by choosing $\alpha = 1, 2, 6$ and for varying sample sizes ($n = 10, 20, 50$).

Here, we utilize the results of Leisenring and Ryan (1992) and compute the distribution of RfD using (14.2.5). For computational purposes, in order to select reasonable values for μ and α, we used the results of published studies on various uncertainty factors. Typically, four to five uncertainty factors are used to determine the RfD. Gaylor and Kodell (2000) provide a table of median values and standard deviations of the logarithms of these uncertainty factors. Their values are based on several independent studies on different databases by various

TABLE 14.5.1 Percentiles ($\times 10^2$) of the RfD distribution $G(r)$

Parameters			$n = 10$			$n = 20$			$n = 50$		
γ_0	γ_1	α	5th	10th	20th	5th	10th	20th	5th	10th	20th
0.0	0.32	1	.0017	.026	.18	.0011	.014	.09	.0002	.004	.03
0.0	0.32	2	.0060	.042	.24	.0039	.025	.13	.00029	.015	.04
0.0	0.32	6	.0132	.063	.31	.0080	.035	.16	.0061	.025	.11
0.03	0.32	1	.0069	.046	.27	.0042	.027	.15	.0024	.015	.07
0.03	0.32	2	.0087	.058	.32	.0051	.033	.17	.0024	.015	.11
0.03	0.32	6	.0153	.073	.37	.0091	.041	.19	.0062	.026	.12
0.10	0.32	1	0.100	.071	.41	.0058	.039	.022	.0034	.010	.10
0.10	0.32	2	.0111	.095	.42	.0069	.044	.24	.0041	.024	.12
0.10	0.32	6	.0196	.095	.48	.0120	.056	.27	.0069	.028	.13

authors. Using these values, we have approximately

$$\mu = \log 3.5 + \log 1.7 = 1.7834$$

and

$$\sigma = \sqrt{(1.64)^2 + (1.66)^2 + (0.60)^2 + (1.72)^2} = 2.96$$

which are used in the computation of $G(.)$. Table 14.5.1 provides selected percentiles of $G(.)$ for different parameter values and varying sample sizes. Figures 14.5.1 to 14.5.3 depict comparative graphs of $G(r)$ for different dose-response shapes and varying sample sizes. The graphs clearly demonstrate that larger sample sizes lead to higher probability of lower RfD values. This in turn means that there is potentially much higher risk in smaller sample sizes. Also, as the shape of the true dose-response relation changes, i.e., when α changes from 1 to 6, we see that RfD can vary quite significantly. This fact also emphasizes the dependence of RfD on the assumed dose-response model in the study.

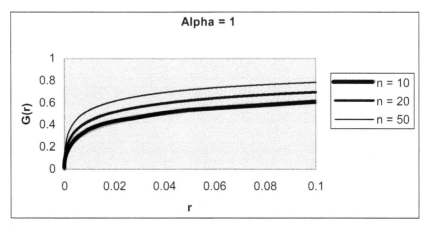

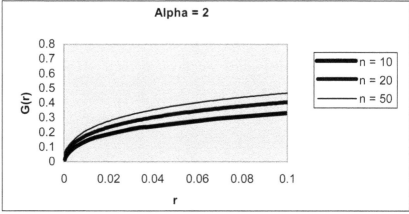

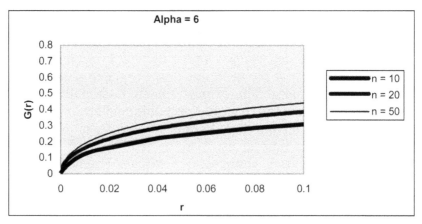

Figure 14.5.1   RfD distribution function $\gamma_1 = 0$, $\gamma_2 = 0.35$.

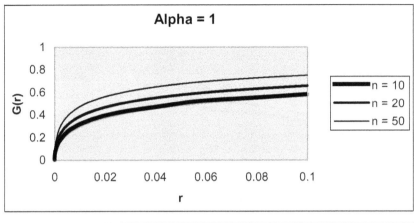

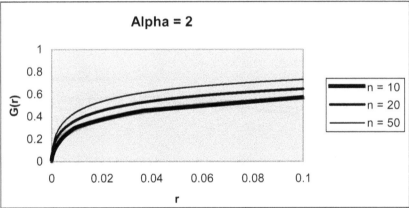

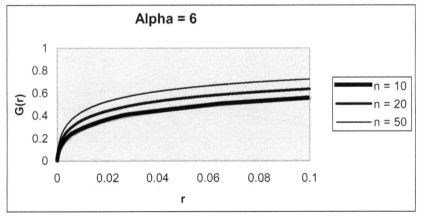

Figure 14.5.2   RfD distribution function $\gamma_1 = 0.03$, $\gamma_2 = 0.35$.

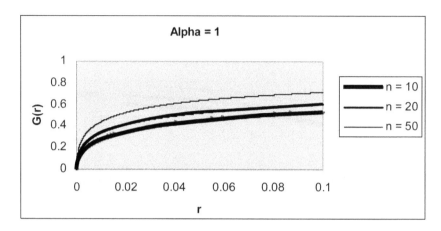

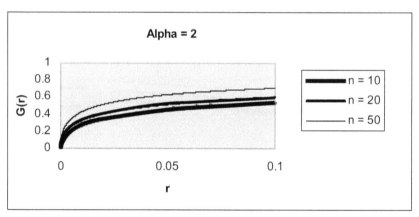

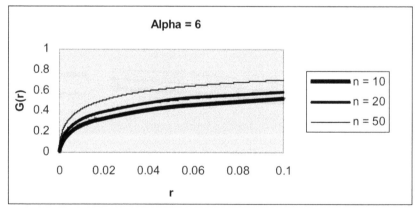

Figure 14.5.3 RfD distribution function $\gamma_1 = 0.10$, $\gamma_2 = 0.35$.

14.6   DISCUSSION

Human health risk assessment is a complicated process that requires the contribution and collaboration of experts in many different disciplines. Because of the high sensitivity of the human body to exogenous agents, the subject of risk assessment has been the focus of many investigations. Regulatory agencies have developed risk assessment guidelines to obtain safe exposure levels for the use of chemical compounds for industrial and commercial purposes. Although these guidelines have brought about major achievements in the regulation and use of toxic substances, at the same time they have raised many interesting and challenging research problems. For example, the risk assessment procedure for all noncancer chemicals relies on the default NOAEL/UF approach. Generally, four to five uncertainty factors are most commonly used. These factors account for uncertainty with respect to sensitivities within the human population, a dose reduction due to using animal data for human effects, uncertainty of extrapolation from sub-chronic studies to chronic effects, uncertainty of extrapolation from a low-risk level to a negligible risk level, and sometimes an additional factor to account for other uncertainties. Recent studies have suggested the use of the lognormal distribution to express the behavior of each uncertainty factor. Using this fact, we have shown, by deriving the distribution of the RfD, that unfortunately risk assessment procedures based on the RfD are not stable and depend much on the experimental conditions. More specifically, we have seen how the properties of the RfD distribution are directly linked to those of the NOAEL. Also, the RfD distribution depends on the sample size, i.e., number of animals used per experimental dose group. Moreover, the RfD distribution changes as the shape of the assumed dose-response model is altered. It is believed that the instability of the RfD distribution stems from the high sensitivity of the NOAEL to the experimental design. Properties of the NOAEL have been investigated by several authors and its usage has been criticized because of its subjectivity and its dependence on the sample size, variability from experiment to experiment, and the dose-response curve. Often, the ratio of NOAELs is used to measure relative toxicity of chemicals. Brand, Rhomberg, and Evans (1999) show that because of the poor statistical properties of the NOAEL, their ratios can contain systematic errors and hence be misleading.

Alternatively, studies have suggested the use of a benchmark dose as an alternative method for risk assessment. The Benchmark Dose (BD) is defined as the effective dose of a substance that corresponds to a small negligible risk. In this approach, the dose effect is described by a mathematical dose-response relationship and a lower confidence limit is calculated for a fixed small negligible value of risk. The BD methodology is finding a widespread popularity because of its stability and its statistical properties. Gaylor (1991) compares the properties of the RfD based on the NOAELs and the BDs using bioassay data from three different sources. His conclusion is that RfDs derived from NOAELs are inconsistent and unreliable. The most important problem in using the BD approach, however, is in developing and utilizing the most adequate model in any specific setting. As pointed out by Williams and Ryan (1997), one of the major scientific challenges for risk assessors is development of appropriate dose-response models. Such models should consider the complex biological processes that cause toxicity in humans. The process of toxicity represents the culmination of several pharmacokinetic, biochemical, and physiological events once the chemical substance enters the human body. To reduce these uncertainties, one approach would be to consider mathematical models that are based on biological principles and incorporate biological and mechanistic information into the dose-response characterization [Andersen et al. (1992)]. As pointed out by Shuey et al. (1994), the goal of biologically based dose-response (BBDR) modeling is a mechanistic description in mathematical form of the sequence of casual events that intervene between administered dose and adverse outcome. Although some attempts have been made to describe the process of building BBDR in certain cases [see, e.g., Gaylor and Razzaghi (1992)], an enormous amount of basic research is still needed in the derivation of quantitative biologically based models.

REFERENCES

Andersen, M. E., Krishnan, K., Conolly, R. B. and McClellan, R. O. (1992). Mechanistic toxicology research and biologically based modeling: Partners for improving risk assessment. *CIIT Act.*, **12**, 1–7.

Brand, K. P., Rhomberg, L. and Evans, J. S. (1999). Estimating non-cancer uncertainty factors: Are ratios of NOAELS informative? *Risk Analysis*, **19**, 295–308.

Dourson, M. L., Felter, S. B. and Robinson, D. (1996). Evolution of science based uncertainty factors in noncancer risk assessment. *Regulatory Toxicology and Pharmacology*, **24**, 108–120.

Flaten, T. P. (1997). Neurotoxic metals in the environment: Some general aspects. In *Minerals and Metal Neurotoxicology* (Eds., M. Yasui, M. J. Strong, K. Ota, et al.), pp. 17–25. CRC Press, Boca Raton, FL.

Gaylor, D. W. (1991). Comparison of the properties of reference doses based on the NOAEL and benchmark doses. *Proceedings of the 84th Annual Meeting of the Air and Waste Management Association*, Paper Number 91–173.6.

Gaylor, D. W. (1994). Biostatistical approaches to low-level exposures. In *Biological Effects of Low Level Exposures to Chemicals and Radiation*, pp. 87–98. CRC Press, Lewis Publishers, Boca Raton, FL.

Gaylor, D. W. and Kodell, R. L. (2000). Percentiles of the product of uncertainty factors for establishing probabilistic reference doses. *Risk Analysis*, **20**, 245–250.

Gaylor, D. W. and Razzaghi, M. (1992). Process of building biologically based dose-response models for developmental defects. *Teratology*, **38**, 389–391.

Hattis, D. (1998). Strategies for assessing human variability in susceptibility and using variability to infer human risks. In *Human Variability in Response to Chemical Exposures: Measures, Modeling and Risk Assessment* (Eds., D. Neumann and C. Kimmel), pp. 27–57. ILSI Press, Washington, DC.

Johnson, N. L., Kotz, S. and Balakrishnan, N. (1994). *Continuous Univariate Distributions—Vol. 1*, Second edition. John Wiley & Sons, New York.

Kodell, R. L. and Gaylor, D. W. (1999). Combining uncertainty factors in deriving human exposure levels of noncarcinogenic toxicants. In "Uncertainty in the Risk Assessment of Environmental and Occupational Hazards." *Annals of the New York Academy of Sciences*, **895**, 188–195.

Leisenring, W. and Ryan, L. (1992). Statistical properties of the NOAEL. *Regulatory Toxicology and Pharmacology*, **15**, 161–171.

Rohatgi, V. K. (1984). *Statistical Inference*. John Wiley & Sons, New York.

Shuey, D. L., Lau, C., Logsdon, T. R., Zucker, R. M., Elstein, K. H., Narotsky, M. G., Setzer, R. W., Kavlock, R. J. and Rogers, J. M. (1994). Biologically based dose-response modeling in developmental toxicology: Biochemical and cellular sequelae of 5-fluorouracil exposure in the developing rat. *Toxicology and Applied Pharmacology*, **126**, 129–144.

Sontag, J., Page, N. P. and Safiotti, U. (1976). Guidelines for carcinogen bioassay in small rodents. *DHHS publication (NIH)*, pp. 76–801. National Cancer Institute, Bethesda, MD.

Williams, P. L. and Ryan, L. M. (1997). Dose-response models for developmental toxicology. In *Handbook of Developmental Toxicology* (Ed., R. Hood). CRC Press, Boca Raton, FL.

Subject Index